# Basic Construction Estimating

# Basic Construction Estimating

JACK R. LEWIS, Ph.D.

*Professor-Architect*
*California Polytechnic State University*
*San Luis Obispo, California*

PRENTICE-HALL, INC., Englewood Cliffs, New Jersey 07632

*Library of Congress Cataloging in Publication Data*

Lewis, Jack R., 1911-
    Basic construction estimating.

    Bibliography: p.
    Includes index.
    1. Building—Estimates.  I.  Title.
TH435.L68 1983        692'.5        82-12378
ISBN 0-13-058313-8

Editorial/production supervision and interior design: BARBARA BERNSTEIN
Manufacturing buyer: ANTHONY CARUSO

Printed in the United States of America

10  9  8  7  6  5  4  3  2

ISBN 0-13-058313-8

Prentice-Hall International, Inc., *London*
Prentice-Hall of Australia Pty. Limited, *Sydney*
Prentice-Hall of Canada Ltd., *Toronto*
Prentice-Hall of India Private Limited, *New Delhi*
Prentice-Hall of Japan, Inc., *Tokyo*
Prentice-Hall of Southeast Asia Pte. Ltd., *Singapore*
Whitehall Books Limited, *Wellington, New Zealand*

# Contents

# 3 SITE WORK 15

# 4 CONCRETE AND MASONRY 25

# 5 WOOD CONSTRUCTION 40

# 6 ROOFING AND WATERPROOFING   59

# 7 INSULATION AND ACOUSTICS   76

# 8 DOORS, WINDOWS, AND GLASS   85

# 9 FLOORS   96

## 17  COMMERCIAL ESTIMATING   164

## GLOSSARY   168

## APPENDIX   180

## PRACTICE PLANS   193

## INDEX   199

# Preface

Almost everything that people do, produce, sell, trade, or accomplish has a cost which must be considered. Some costs are measured in time spent; some costs are a convenience or inconvenience; some costs are in relations with other people, but most costs are in amounts of money involved. Every time a person purchases a product there is a consideration for the cost of that product measured against its value or use to the buyer.

The manufacturer of any product must consider a number of costs: those of the actual material required; of labor to produce a marketable item; of operating a business; of tools, equipment, permits, licenses, and insurance; selling or advertising costs; and of course a profit. Very few business ventures can stay in operation for very long if they do not calculate all of these costs and somehow keep the selling price of their product in line with the price of their competition. This process has developed a procedure that in building construction has been misnamed an "estimate."

The dictionary states that an estimate is "an approximate opinion, a rough calculation, an evaluation." None of these definitions really applies to construction cost estimating in the fullest sense. It is true that preliminary estimates are educated "guesstimates," usually based on reliable knowledge, that are used to determine **approximate** costs for budget purposes, insurance, loans, and other items. It is also true that the estimate given by a contractor, subcontractor, or material supplier is not an estimate but a hard-cost figure rather than an approximation. In order to price the cost of construction of a building the builder needs an accurate take-off of quantities of materials; cost of labor to install that material properly; adequate knowledge of costs for permits, inspections, tests, security, and insurance; and general office and job-site costs. After all of these are known, the builder must then determine his profit (or loss) to ascertain if he should build the building.

Estimates for building construction are not limited to those needed by the actual builders. Architects and engineers need estimates to provide information to clients; lending agencies loan money on estimates of property value; fees for permits to build are based on construction cost estimates; and almost every other phase of endeavor connected with building is in someway connected to an estimate. The purpose of this book is to try to show how a reliable construction cost estimate may be made by using a few rules, some common sense, elementary arithmetic, and a little knowledge of construction. The illustrations and problems are for residential construction and do not attempt to discuss design, materials, or related matters. These subjects are adequately covered by the publications listed in the Appendix.

In writing this book the author has done so with the full realization that it is not a complete book to qualify the reader as an estimator. Many people in many fields try to estimate the cost of construction, each from his or her own position and with a different end result that they hope to achieve. This book is intended as an introduction to estimating practice with the knowledge that every detail of every type of construction, even of every type or site of residential construction, is not covered extensively and completely.

The person who expects this book magically to give him the complete answer to construction costs will be critical of the fact that cost figures given in this text are not accurate for a particular area or time. This is freely admitted. All cost figures for material or labor or both are to be used to **practice** estimating. This book is written as a text to **teach** estimating practice, not as a handy reference for current costs. Current costs must be obtained by making calls to local dealers or subcontractors or by careful and intelligent use of annually updated references noted in Chapter 2.

The second difference between the information contained in this book and the methods used by most general contractors or owner-builders is the fact that a great deal of actual estimating and subsequent construction is done by subcontractors specializing in various trades. This means that the prime contractor will probably fully estimate only that portion of the complete project that he expects to do with his own crew and will take subbids for the remainder of the work. In some cases the prime contractor may "guesstimate" the cost of subtrades work to provide some guidance in evaluating the subcontractors bids, but there is generally neither the time nor the need to estimate every trade fully and accurately. Each subcontractor is better qualified to determine his own material costs, labor costs, overhead, profit, and time schedule.

Professional estimating is not for all people. In addition to a knowledge of construction, the estimator must possess an interest in detail and a feeling for thoroughness. Many small items that are necessary for complete construction are not shown on drawings or called out in the specifications, so the estimator must be able to include these in his take-off. An extensive knowledge of mathematics is not required but the estimator needs to be able to visualize similar areas or to associate various shapes that may be combined. Perhaps the most important quality for an estimator to possess is the ability to keep up to date with costs of material and labor as they pertain to the location in which the work will be done.

JACK R. LEWIS

# Basic Construction Estimating

# chapter

# 1

# Construction Documents

Before any calculations for estimating are begun, the estimator must have a good working knowledge of construction as well as a good knowledge of reading construction drawings and their accompanying specifications. The drawings must show *quantities* via plans, sections, details, and diagrams. The specifications indicate the *quality* of the material to be used and are usually in book form; however, quite often the specifications for residential work are abbreviated to the extent that they appear as notes on the drawings. The drawings and the specifications complement each other and what is indicated in one document is considered to be indicated in both. The two documents, plans and specifications, taken together are known as the *construction documents* or *contract documents*.

## 1.1 BLUEPRINTS

The drawings are most often referred to as "blueprints"; however, this is not necessarily an accurate description. Drawings or plans, elevations, sections, and details are first drawn on transparent linen cloth, treated paper, or Mylar plastic. These drawings are then reproduced in multiple sets on chemically treated paper. In this reproduction process natural or artificial light is used to expose the drawings much like a photographic process and the resulting "print" is identical with the original drawing. In earlier times the exposed final print turned out with white lines on a blue background, hence the name "blueprint." Although some blueprinting is still done, most reproduction currently is of blue or black lines on a white background. Despite these changes, reproductions are still referred to as "blueprints" or, more correctly, simply as "prints."

Construction drawings have fairly definite rules for indicating walls, doors, windows, materials, connection of details, and dimensions to show the sizes of various shapes chosen. It is not possible in this chapter to give all methods or the requirements for all situations in drafting. However, a quick review plus a few hints may be of help in understanding the meaning of lines, figures, symbols, and their relation to each other.

Figure 1-1 shows the correct way in which some symbols appear on the floor plans. Some variation in the design of each may be expected, but usually these variations are not extensive enough to confuse their meaning. Sizes or, more properly, dimensions are shown to control the location of parts shown as symbols, to indicate length, width, or height of floors, elevations, sections, and details. Again, there are rules or proper methods of dimensioning. Basically, all dimensions are shown to the face of *structure,* not to the face of finish material, as this surface varies with the thickness of the material used. Door and window locations are usually indicated to centerline (₵) of the unit from an established wall line. From door and window details the carpenter is then able to properly frame the unit, allowing space for frame, shims, and other parts. Some typical dimensions are shown in Figure 1-2.

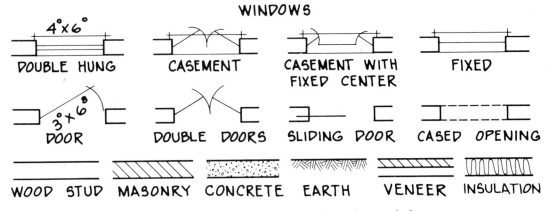

**Figure 1-1**   Floor plan symbols.

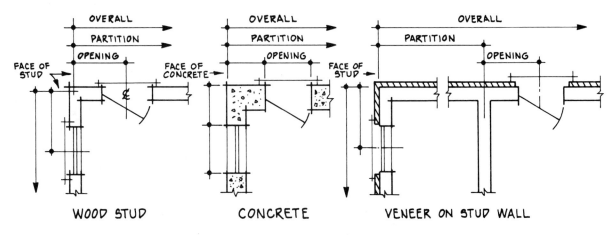

**Figure 1-2**   Floor plan dimensions.

In order to show as many important conditions as possible, sections are shown at various locations. Sections are a visual interpretation of a vertical slice through the construction at the point shown and indicate the position, size, shape, and other detail of the members cut through. Sections are usually shown at a larger scale than the plans or elevations and can therefore show detail more clearly and to better advantage for the viewer. Since it would be practically impossible to show most parts of a building at full size, drawings are made to a *scale*. Scale is a method of showing a drawing at a ratio of its true size and again there are general rules to be observed. Plans (foundation, floor, roof), elevations, and some sections are usually drawn as ¼″ = 1′-0″, or ¼″ on the drawing represents 1′ (one foot) of actual building. This means that a building that will be constructed 50′ wide will be shown as 50 × ¼″ or 12½″ on the drawings. Sections and details are usually drawn at ¾″ = 1′-0″, 1½″ = 1′-0″, or even 3″ = 1′-0″. Occasionally, full-size drawings are made and indicated as "F.S." or 12″ = 1′-0″.

A full set of prints (from drawings) usually also include a *site plan,* which shows the size of the property; its legal description; and the location of the building, driveways, sidewalks, fences, and any other improvement to the property. Landscape work, where required, is also shown on the site plan. Vertical heights (elevations) are noted and often shown by irregular lines showing the same natural or finish elevations throughout the site. The scale of the site plan is usually different from the scale of the architectural drawings and most often is drawn to an engineering scale such as ¼″ = 100′-0″ and fractions are shown as decimals of the whole. A typical site plan is included in the practice drawings included in this book and is discussed further in Chapter 3.

To begin to "read" a print takes a little practice. Probably the first thing is to review all the prints for a general idea of the entire project. This orients the builder before he starts to scrutinize the prints further. After the preliminary review the builder usually will carefully review the floor plans, elevations, sections, and details. This review is to familiarize him still more with the project and during this review the estimator will make notes to indicate the general materials, and possible subcontractor bids, that will be needed. Not all work on a project is "subbed" out, so perhaps the grading, excavation, concrete and its forms, carpentry, and such "buy-outs" as finishing hardware may be done by the principal or prime contractor. Most of the other work required is to be provided by subcontractors who will make an "estimate," and a bid to the prime contractor, for their part of the project.

The estimator needs to be able to visualize the drawings in three dimensions as well as to visualize the actual installation procedure in the field. By so doing he will be able to contemplate the proper step-by-step action of the field crew. He will *also* be able to determine what equipment, scaffolding, power, or other items are required even though they are not shown on the drawings or called out in the specifications.

## 1.2 SPECIFICATIONS

Complementing, or perhaps supplementing, the prints of the drawings are specifications which call out the *quality* requirements for the various materials that are to be used in the construction. In the early planning stage the preliminary specifications simply list the materials in a decisive manner to identify them by established standards such as Federal Specifications (Fed. Spec.), American Society for Testing and Materials (ASTM), manufacturers' association grades, accepted gauges or sizes, and manufacturers' trade names. These preliminary specifications are very useful in helping to determine budgets, materials to be used, and of course the various subcontractors and material suppliers in what materials to purchase and install.

In 1964, the Construction Specifications Institute (CSI), a national organization of architects, engineers, specifications writers, contractors, manufacturers' representatives, and suppliers, recommended a format of 16 divisions for the arrangement of construction specifications. Today the U.S. government, many state governments, and most architect/engineer firms use the format, now called the *Uniform System*. Manufacturers' catalogs now use the 16 divisions to identify their products. With the possible exception of very small residential or owner/builder projects, proper use of the Uniform System simplifies specification organization, and any person who wants or expects to do construction estimating should be familiar with this format. A complete list of the 16 divisions and *possible* sections under each division head are included in the Appendix.

Two methods for preparing specifications are commonly used in residential work. The most common is to hand letter or type abbreviated requirements directly on the drawings. The second method, to produce a "book" that will accompany the prints, is more generally used for projects larger than residential. In the first case the draftsman simply letters information and a proper directive arrow, without additional requirements, for material such as "5/2″ shingles—5½″ exposure." This means that the shingles will be a butt thickness that will provide five shingles in 2″ and that they will be laid with 5½″ exposure from the butt end. They may be cedar, redwood, or other material commonly used for wood shingles and the installer is assumed to know how to place, nail, and otherwise install them properly. Similar indication is used for many other materials and may have some advantages in that the installer is allowed to use his own method (a performance specification) as long as the final use or appearance is proper. When specifications are written in book form or large projects the several sections would have a similar format, which includes scope of the work, related work, handling and storage, materials, testing, samples, models, shop drawings, installation guarantees, and cleanup. All of these requirements are spelled out in detail and most generally do not allow the installer to use his own option to any great extent. Information contained in either the drawings or the specifications should *not* be duplicated in the other document if it can be avoided.

## 1.3 ADDENDA AND CHANGE ORDERS

In many cases of construction the prints of drawings and the specifications are the only construction documents. Changes do occur, however. If a project is in the bidding stage, changes may be made by *addenda,* which are simply clarifications, additions, deletions, changes, explanations, or modifications to the original documents. These are provided to all bidders to allow for any monetary recalculation necessary to include the items changed in the bids. The items called out become the same as though included in the original proposals.

Once a construction contract is signed by the owner and the contractor and construction has begun, all changes should be authorized by *change order.* These are documents stating the changes in the work, cost of the changes, and difference in time of construction (if any), which are signed by both the owner and the contractor. They modify the construction contract and may include almost anything from approval of substituted materials to major changes in the size of the building form, and may be added expense to the construction or a reduction in cost.

## 1.4 PERMITS

Every state or area in the United States has a building code by which various types of construction may be governed. In general, the local city or county authorities adopt appropriate codes and may enact additional ordinances for planning, zoning, fire protection, environmental control, and other requirements. The contract documents for construction will be reviewed, for a fee, by a department or agency and permission granted to construct the project. The project designer is responsible for meeting the requirements of these codes and ordinances and the builder is responsible for constructing the work in accordance with the drawings and specifications. During construction periodic field inspections are usually made to ensure that the work is being done as shown and approved in the construction documents.

## 1.5 SAMPLES AND SHOP DRAWINGS

Either before construction starts or during construction the matter of substitutions and samples of material comes up. Either of these means that the subcontractors or material suppliers must provide samples of the actual material that will be used on the project. In some cases these materials are those that have been originally specified, but they may also be substitutions. The estimator is interested because any substitution may change his costs. After construction has started, various fabricated items may require shop drawings—the drawings made by a subcontractor to indicate the *exact* size, shape, and material that the fabrication will have. Shop drawings may also change the cost if the designer requires different materials or other changes.

# chapter

# 2

# Methods for Estimating

Estimates for construction costs are used for different reasons and so are made by different methods and provide different answers. Some methods are adequate for the purpose intended and are relatively simple. Others are more detailed and require more time but provide a better and more concise answer.

## 2.1 COST PER SQUARE FOOT

Probably the first need for an estimate of construction costs is when the owner or client wants to know *approximately* what a proposed project might cost to build. For this he goes to an architect or engineer or perhaps to a building contractor, from whom he may obtain a *cost-per-square-foot* estimate. This estimate is truly an estimate since it is made on the square-foot area of the proposed building multiplied by a known or assumed cost of construction for each square foot of building. To arrive at the square-foot area is not necessarily difficult; simply calculate the number of square feet of floor surface to include all the area from the *outside* faces of the building, *not* a total of the various finished *interior* spaces. Where a porch or other semicovered construction occurs it is usually considered as only one-half of its actual area. Unfinished areas such as garages, attics, and basements are kept separate, as these also are generally figured at different costs per square foot.

There are two principal difficulties in calculating square-foot cost of construction. First is the matter of floor area measurement. Many drawings at the preliminary stage show only sizes of major rooms and do not include halls, stairwells, and porches, and only occasionally show overall exterior dimensions. Buildings must have walls and the space occupied by them costs the same, or more, than any enclosed space in a room. It is therefore vitally important to know the overall sizes of a structure, and these are normally given from the faces of wood stud wall or the exterior face of concrete or masonry. In residential work, as well as some other types of occupancy, the finish of stucco or wood applied over the wood studs is rarely more than an inch thick and does not affect the overall area.

The second problem with square-foot costs is more serious—the cost in dollars per square foot. This means that in theory every square foot of floor area costs the same to provide, but anyone will easily realize that the open space in the center of a living room does not cost the same as a similar square-foot area at the kitchen cabinets. In using this method, however, every square foot of space is considered to cost the same, and the real problem is to determine the proper cost depending on the type of construction.

Most residential construction is of wood with an exterior covering of stucco or wood, wood floor systems, plaster or drywall interior covering over wood studs, and wood shingle or shake roofs. Considerable western and southern residential construction is built on concrete slabs directly on the ground, thereby eliminating concrete or masonry foundation walls and wood floor framing, but retaining all of the remainder of the wood stud construction and wood roof construction. In many designs there are also concrete or masonry walls, either as solid fire walls or as structural requirements, or perhaps only as thin veneer over wood construction. Roofing may also be different, using wood shakes or shingles, clay tile, slate, built-up asphalt roofing on flat surfaces, or even metal roofs. Of course, the interior trim, cabinet work, mechanical and electrical equipment, and other parts may vary considerably.

These differences are emphasized because square-foot costs must be calculated on *similar kinds* of construction, or ideally, on *identical* construction. A true cost cannot be arrived at by trying to improve different types of construction or by using cost figures from a publication that may not be based on similar work. To arrive at reliable square-foot costs, many firms try to keep their own records. The idea is great—but perhaps not as easy as it would seem to be. The first requirement is a complete and *accurate* description of a project in regard to size, material used, grade of construction, final cost, and of course the cost per square foot. Date of construction, length of time for construction, location (especially in regard to hills, water, landscape, city/suburb, etc.), and any other data may also be helpful. Over a period of time this collection of information may be used to compare similar types of projects, varied by additions or subtractions due to quality of construction, inflation, different location, time of year, availability of materials or labor, and similar items. To be useful, all of this information must of course be *local*.

When the actual square-foot area is finally calculated and a legitimate cost per square foot is found, all that is necessary is to multiply one by the other and *presto!* you have an approximate cost, or a preliminary estimate.

## 2.2 QUANTITY SURVEYS

For a cost estimate that is as close to the actual cost as possible, the quantity survey method is the best approach. This is the method used by contractors, subcontractors, material suppliers, and anyone else who needs or wants a complete cost figure, that is as accurate as possible. The system is simple: from drawings or field surveys the actual quantity of each material is figured, then these quantities are multiplied by the cost of material, cost of labor for each unit, plus overhead and profit.

The working drawings of any project essentially show the quantity of each material through plans, sections, and details. To take off the amount of each quantity requires a bit of experience in reading drawings. Each material needs to be accurately figured from the sizes and shapes shown on the drawings, especially from sections and details. Like materials may be lumped together or held separately. Location and distance aboveground or other working platform levels often involves different labor costs for installation of the same material. Quite a few materials such as concrete forms, rough hardware (nails, bolts, lags), and various pieces of rented equipment necessary to construct the work will not be shown on any drawing so will have to be anticipated and figured by the estimator. Some items that are not shown on the drawings may be called out in the specifications. Labor costs of installation, permit costs, testing costs, overhead, and profit will not be shown in either place. All must be included in arriving at the true cost of construction.

## 2.3 LUMP-SUM AMOUNTS

Often, the architect, contractor, or others interested in construction costs either do not have the time or the opportunity to do a square-foot estimate or a quantity survey. Incomplete projects, changes, or other factors may make a total cost more convenient. Such costs are lump sums which include the costs of materials, labor, overhead, profit, and any other items that affect the in-place cost. Almost invariably this type of estimate is used on construction change orders when there is an addition or deduction during construction, and it is often used as part of other types of estimates. Lump-sum or in-place estimates of necessity use percentages or some other part of such items as overhead, profit, and other indeterminant items. This type of *estimate* is certainly convenient, could be fairly accurate, and in many cases is all that is required.

## 2.4 UNIT COSTS

The unit form of cost estimating is normally used only in certain types of construction. Unit costs may be used to determine costs for additional yards of concrete, additional lengths of piling, amounts of earth cuts or fill, and similar work. A great deal of civil engineering work is constructed on the basis of unit costs. In most cases this form of cost engineering will not be used in residential work except perhaps for excavation work where there may be a question regarding the extent of rock substrate or other suspect subsurface conditions.

**2.5 APPRAISALS**  Appraisals are normally considered to be outside the scope of construction cost estimating since such methods are used only after a project has been constructed. Appraisals are made by those experienced in estimating existing projects based on the market value of the project, surrounding property values, replacement or repair involved, and in many cases as collateral for real estate loans. Such appraisals are based on square-foot cost to replace, and are seldom arrived at by quantity survey since there is a very limited opportunity to determine amounts accurately, as many are hidden by existing construction. Real estate agents, lending agencies, some city departments, and perhaps some prospective clients for some projects may use appraisals, but this is not construction estimating.

**2.6 MATERIAL**  The actual *estimate* put together by a contractor is a combination of several items: materials, labor, overhead, and profit. The material, via quantity survey, is probably the easiest to calculate. It is simply the amount of a material it takes to provide the project, plus some variable for waste. The amount used as *waste* may be zero in cases of mechanical units or may be as high as 30 to 40%, depending on unit sizes obtainable, cutting requirements, minimum ordering quantities, or a design that requires an excess of odd shapes or sizes. Some waste may be reduced by careful field control or may be minimized by long-range planning. Extra concrete is difficult to save, but if used to make pier footing blocks that may be used on another job is a small but definite savings. Careful cutting of lumber may result in a savings by using short lengths for blocking. Sometimes, however, an assumed material cost savings may actually be a loss due to the labor involved. This is often true when labor is used to pull nails from lumber and stack it. In this case many contractors find that the value of the used lumber, plus the labor cost to clean and stack it, is more than new lumber costs, so often burn or otherwise dispose of once-used material.

**2.7 LABOR**  Probably the most difficult part of cost estimating is that concerned with labor required to cut, fit, or install the various materials used. Several factors influence labor costs. The most apparent is that of human ability; some mechanics simply comprehend more readily and work more rapidly and efficiently than others. There are few rules that spell out the speed or productivity of craftsmen and virtually no rules to produce installations in "a workmanlike manner." Specification of "first-class work" means one thing to one person and something entirely different to another person. Some unions have rules which tend to set maximum amounts of work or units that a member may install in a day, or that require a given number of "tenders" or helpers to assist each journeyman or group.

There are several established organizations that provide annually updated information regarding labor for installation of most types of construction.[1] These guides are valuable only as general information since many factors, such as individual ability, location, type of structure, weather, union rules, and similar items are not easily included. For the established contractor, with a reasonably good time-keeping system, the best source for labor costs are his own time records. The experienced foreman or superintendent will be able to assign labor time and costs properly to each of the various items of work accomplished, which in turn may serve as a basis for estimating labor time and costs on a future project. On very large commercial-industrial projects a timekeeper may handle this information and organize it for later use. Crew sizes and the ability of crew members to work together efficiently also greatly affect some types of labor costs. Labor and material costs, or in-place costs, included in this book are for general guidance and illustration only and should be checked for local variation and changes over time.

## 2.8 OVERHEAD COSTS

Strictly speaking, overhead costs are not a method, but are a necessary part of any construction estimate. There are two principal types of overhead: *permanent* or *fixed items* relating to the cost of maintaining a business, and *variable* costs that fluctuate with each project. Both of these must be considered when figuring the total cost of construction.

Every business has permanent or fixed costs. These are the costs that go on regardless of the type or location of work and are affected only slightly by the amount of work. They are costs that cannot easily be charged to any one project and include office rent, utilities, office equipment, storage or assembly space, stationery, insurance, and many similar items. Permits or registrations to do business are fixed costs, but permits for each project are not. Salaries of office personnel are usually a fixed expense, as it would be quite difficult for a secretary to keep accurate records of the amount of time it took to type a letter or answer a phone call. The in-house estimator may be able to log his time against a project if he only worked on one at a time, but is often concerned with several at once and may be in somewhat the same position as the secretary. Certainly, "the boss" would have difficulty in accurately keeping track of his time, as he may be engaged in scouting for more work, inspecting a going project, attending an association meeting, talking to a client, or any one of a dozen things. Total costs for these fixed overhead expenses may be calculated on a yearly basis and an hourly or weekly amount charged to each project or equated on a cost-of-construction percentage.

Variable costs are those directly related to each project and may be lump-sum amounts or, in some cases, a percentage of cost. Included in this category are move-on and time costs for a job "shack," utility costs for the site, fencing and security measures, job

[1]Robt. Snow Means Co., Inc., Kingston, MA 02364; Richardson Engineering Services, Escondido, CA 92025; and Lee Saylor, Inc., Walnut Creek, CA 94596.

permits, rental costs of sanitary toilets, insurance and bonds that are required, and all other items applied directly to a particular project. If the job *shack* is a permanently built movable unit such as a trailer, or if some expensive equipment is owned by the contractor and used on the project, a fair per-day rate based on original cost, depreciation, and rental rates should be charged to the job. Some items, such as temporary fencing or storage sheds, may be removed at the termination of the project and a salvage value or per-day cost should be used. Construction supervisory personnel are usually charged as a project expense, but if they handle more than one project at a time, a cost commensurate with their time should be used.

## 2.9  PROFIT

Every business expects to make a profit and the contractor is no exception. How much profit is the question and each project and each contractor will require a different answer. It should be rather obvious that if a business cannot return a profit on the money invested which is greater than the interest that may be obtained by putting that money in a bank, there should not be a business investment. Therefore, we can safely assume that an interest or profit return of not more than 7 to 10% is not adequate. The *markup* for overhead and profit in many retail stores is from 33 to 50% over wholesale costs, so if the project overhead figures at 20.9%,[2] the 1979 current average, perhaps a profit of 15 to 25% is not greatly out of line.

As in most any other business, the construction industry and the individual contractor have times of plenty and times of scarcity. If there is plenty of work the contractor may feel that he can obtain a higher profit, but in periods of slowdown he may take work at nearly cost in order to keep his business known and his trained and coordinated crews together. Even more than in most other types of businesses, the profit factor varies in construction. The costs for labor and materials must be estimated well in advance of actual purchase, and quite often these costs increase greatly before the project is started or completed, adversely affecting the profit figure.

## 2.10  THE "TAKE-OFF"

Proper estimating requires a definite systematic approach, one that will not only give correct answers now but will also be useful during construction or for future reference. Most estimators use multicolumn pads to indicate item unit, quantity, cost of material, labor, total material and labor (in-place costs), and perhaps a space for special notes or identification. Figure 2-1 shows a typical estimating sheet. The estimating sheet contains a great deal of information and the estimator may often include quick sketches, notes, subcontractor or material suppliers names, instruction to field superintendents, and other useful information. One of the most common faults found in some estimating sheets is lack of the identification as to where the

[2]Robt. Snow Means Co., Inc., *Building Construction Cost Data,* 1979, p. 286.

various materials/labor are to go on the project. This may mean that the field crew uses wrong lengths of lumber or installs material in different quantities, resulting in shortages that might have been avoided. Estimating and its variations are the same for the subcontractors as for the general contractor, the most common difference being in the amount or variety of materials considered.

BUILDING  Residence - E.L. Brown
LOCATION  Anytown  Lot 1 BKK

ARCHITECT  Lewis                    SHEET # ___1___ of _10_____
ESTIMATOR  Smith  ✓ OR H.S.   DATE  ___2 Febr. 19 -_____

| MATERIAL | SIZE | UNIT | QUANTITY | UNIT PRICE | TOTAL MATERIAL | LABOR | TOTAL |
|---|---|---|---|---|---|---|---|
| CLEARING | | | | | | | |
| Trees - removal | 6" | | 6 | 50 - | — | 300 - | 300 - |
| Grade level | | Estimated | | | | 200 - | 200 - |
| | | | | | | | |
| EXCAVATION | | | | | | | |
| Ditch ftg. | 24" | lin. ft. | 160 - | 5 - | — | 800 - | 800 - |
| Machine rental | — | — | — | — | — | 100 - | 100 - |
| Hand clean | | Cu. yd. | 20 - | 3 - | — | 60 - | 60 - |
| | | | | | | | |
| Forms | | | | | | | |
| Plywood | 4x8 | Sq. ft. | 150 - | .50 | 75 - | 150 | 225 - |
| Labor | | | | | | | |

**Figure 2-1**   Typical estimating sheet.

# chapter
# 3

# Site
# Work

Every project must have a site on which it is to be constructed. This site may be a river and its adjacent banks in the case of a bridge, or a long narrow strip of land for construction of a road, but more commonly the site is a piece of land bounded by roads or streets and probably near required utilities. Land in cities or counties is usually zoned for use as residential, commercial, industrial, or other occupancy and distances from property lines are established by codes or local ordinances. Construction materials that may be used depend on the type of occupancy, fire district, and various other local requirements.

## 3.1 SITE LOCATION

Location of land sites is accomplished by a survey description, indicating the exact location of the parcel on the earth from monuments or "bench marks" established by governments. In the United States, a system of grids and base lines, located for the most part by the U.S. Coast and Geodedic Survey, are used to identify survey points. From these permanent points a licensed surveyor may use instruments to establish the corners of any piece of property in this country. Most pieces of property are located in cities by a lot number within a numbered block, together with a notation indicating the general location within an original city plat or an "addition" or tract. Figure 3-1 shows the location and legal description of a possible city lot.

## 3.2 SITE WORK

Site work includes a wide variety of construction, often designed or installed by persons with very different talents. Existing trees, brush,

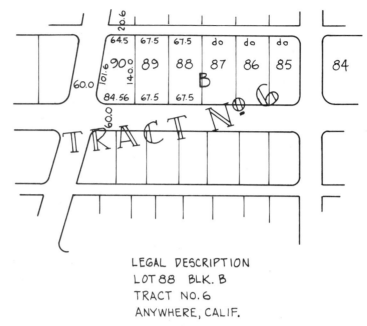

LEGAL DESCRIPTION
LOT 88 BLK. B
TRACT NO. 6
ANYWHERE, CALIF.

**Figure 3-1** Lot location and legal description.

old buildings, and buried utilities may have to be removed. Hilly land may have to be leveled, requiring various amounts of cut or fill. Landscape work, including plantings, sprinkler systems, walls, fences, and water drainage of different types may be included. Sewer, water, and perhaps buried electrical service must be installed and in cases of some commercial work, railroad track, wharfs and docks, roads, and bridges are a part of *site work*. Probably the most familiar site work, however, is excavation and grading to level the building site and provide for footings and basements.

## 3.3 SITE CLEARING

Unless the site is entirely free of vegetation, and most sites are not, someone must clear off existing trees, brush, weeds, and other growth that may be detrimental to construction of the project. This is work for a bulldozer, chain saws, and perhaps some excavating equipment to remove roots and stumps. Removal of old fences, old walks or pavement, existing buildings, and "appertenances" must also be accomplished. For the most part this work is done with common labor or by a subcontractor who specializes in this work, and costs vary widely. In general an estimator may divide this work into several categories: demolition, tree removal, clearing and grubbing, and rough grading. Disposal, on-site or off-site, is also a problem to be considered. Demolition of buildings is normally estimated by the square-foot floor area of the building and runs from about $500 for a 1500-sq ft frame building to about $1500 for a concrete structure. Paving is usually ripped and hauled away and should be figured at $5.50 to $6.00 per cubic yard based upon 3″ thickness. Complete removal of trees, including roots and stump, varies with the trunk size but should be removed for $50 for 6″–8″ trunk size and about $200 to $250 for 24″ trunk size. Most of this demolition has little or no salvage value, so may be a straight cost.

**3.4 GRADING**

Unevenness of the land is shown by *contours* established by the surveyor. These contours show the elevation and vertical distance above the sea or bench mark, and are the same all along a contour line. Unless contour lines are adequately marked in sequence, there is virtually no way that a reader can discriminate between a hill and a depression. Figure 3–2a indicates a site with a hill, and using the same contour lines, Figure 3–2b indicates a site with a depression in the same area.

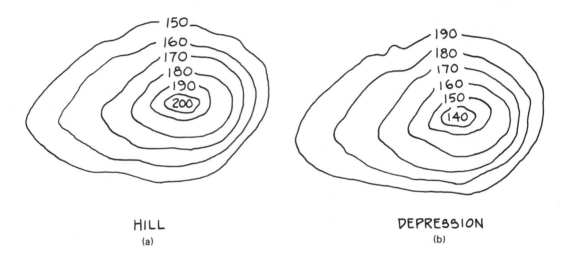

HILL
(a)

DEPRESSION
(b)

**Figure 3- 2**  Site contours.

Grading may be very limited or may be quite extensive, depending on the countours of the site and the required finish building platform. Earthmoving is determined in cubic yards (1 cu yd = 27 cu ft), and is either cut or fill to meet the elevations on grade stakes established by the surveyor or an experienced grading contractor. Calculations for amounts of earth moved may be done by using a series of cross sections taken through the site, or by calculating average elevations on a grid pattern and adding and subtracting cut or fill. The latter method is used in estimating quantities in Figure 3–3, and may be used on any site.

In this problem the question was the exact elevation of the site with a balance of cut and fill. If finish grades are given, and they usually are, the same method may be used with the exact finish elevation know and perhaps with some earth to be disposed of or brought in as fill.

**3.5 EXCAVATION AND BACKFILL**

"Excavation" includes removal of material for basements, footings, utility trenches, and similar locations. The required removal may be made in solid rock, sand, gravel, earth, or any combination. Backfill

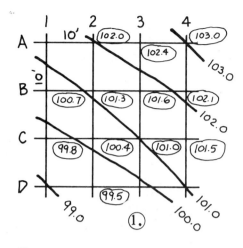

| | Average Elevation | Above 101.125 Cut | Below 101.125 Fill |
|---|---|---|---|
| | (3.) | | (4.) |
| AB12 | 101.375 | 25.0 | |
| BC12 | 100.55 | | 57.5 |
| CD12 | 99.675 | | 145.0 |
| AB23 | 101.825 | 70.0 | |
| BC23 | 101.075 | | 5.0 |
| CD23 | 100.225 | | 90.0 |
| AB34 | 102.275 | 115.0 | |
| PC34 | 101.55 | 42.5 | |
| CD34 | 100.875 | | 25.0 |
| | | 252.5 | (5.) 322.5 |

(2.) Ave. plot for balance cut/fill.

101.5
99.0
103.0     $\dfrac{404.5}{4} = 101.125'$
101.0
404.5

(6.)
$\begin{array}{r} 322.5 \\ -\ 252.5 \\ \hline \div\ \dfrac{70.0}{900} = .077 \end{array}$     $\dfrac{101.125}{\ } + \dfrac{.077}{101.202'}$ Elev.

1. Estimate elevation of grid corners from contours.
2. Determine approx. finish grade by corner average of plot.
3. Average elevation each grid square.
4. Add or subtract from 101.125 for cut or fill and multiply by area (100 sq ft).
5. Add columns for total cut or fill.
6. To complete balance, find difference between cut and fill divided by total area. Add fraction to approx. elev. 101.125′.

**Figure 3-3** Grading-quantity estimation.

is the replacement of material around foundations, footings, beneath slabs, against basement walls, and in similar areas. Each of these operations involves the use of shovels, backhoes, bulldozers, and similar equipment to remove, backfill, compact, or just to move the excavated material. In most cases there is an excess of excavated material over that which may be used for backfill, and this excess is used to fill the site to desired grade elevations or is hauled off-site to a convenient dump. Compaction of backfill is done by jetting or flooding with water, or by solidifying with rollers or hand-operated air-powered compactors. Most backfill is compacted from 85 to 95% of original density in accordance with ASTM Standard D1557.

In many conditions, where the earth is solid enough, no excavation is required to accommodate concrete forms. This means that excavation for footings is made the exact size so that concrete may be placed against the earth without forms. Where forms are required, as is true with most basement construction, the excavation must be made large enough to allow for installation and removal of the forms. There is no problem with the interior face of the wall, as it is accessible from the basement excavation, but the exterior face will require at least 18″ more excavation beyond the wall line to allow for forms. Rock excavation will probably allow almost vertical excavation, but excavation in other materials may slide or cave in, so that the width of excavations at the top is usually wider than at the bottom. This difference is required by "angle of repose", the angle at which it is assumed that earth, sand, or gravel will remain in position without sliding. Excavation in compacted sand and gravel or good solid earth is made on a basis of 1 (foot) vertical to 1½ horizontal or 1:1½; in sand this slope changes to one vertical to two horizontal or 1:2, or more (Figure 3-4).

**Figure 3-4**   Angles of repose for excavation.

Excavation is made and estimated in cubic yards (27 cu ft = 1 cu yd). Table 3-1 indicates some quantities that may be helpful in estimating excavation using different methods or equipment. As stated before, these are averages and may vary widely depending on equipment and personnel using that equipment.

**Table 3-1**   Estimating excavation

| *Excavation* | *Average cu yd per 8-hr day* |
|---|---|
| Hand shoveling and truck loading | 8–10 |
| Tractor shovel, 1 cu yd/bucket | 350 |
| Backhoe, ¾ cu yd/bucket | 75 |
| Ditching machine, 18″ × 5′0″ deep | 350 lin ft/hr |
| *Backfilling* | |
| Shoveling, no compaction | 12–20 |
| Bulldozer or tractor shovel | 550 |
| Sheepfoot roller, 5′0″ wide at 2½ mph | 500 |
| Pneumatic tamper, man-operated | 30–40 |

Even with the use of machines to do most of the excavating work there is usually a considerable amount of hand trimming required to clean up corners, edges, and small areas not accessible to machines. There is also a "swell factor" inherent in excavation work, which means that 1 cu yd of excavated solid material may measure approximately 20% more in sand and gravel, 30% more in normal earth, and as much as 50% more when rock is removed. The *swell factor* affects the amount of waste material or backfill that must be removed or used on the site. Trucking of excess material to dumps also needs to be considered where applicable, and the average time required to load, drive, dump, and reposition a 10- to 12-cu yd truck for a 1-mile haul distance is about 30 minutes. Time of haul, *each way,* needs to be added to distances greater than 1 mile at a rate of 40 mph.

**3.6 OTHER FOUNDATIONS**

Residential work is usually installed with continuous exterior foundation walls on ribbon or pad concrete footings. The actual wall or stem may be poured concrete or masonry, and in many cases may be designed as retaining walls to hold earth embankments. Occasionally, however, there are conditions that require other types of support. Unstable soil near beaches, rivers, or marshland may require pile driving, and noncompacted soil or poor natural soil may require cassions. Each of these operations is expensive and requires the help of experienced contractors.

Pile driving is simply the driving of wood, steel, or concrete "piles" shaped as poles into the ground to a sufficient depth that the friction generated along the length of the pile will provide a required bearing capacity to sustain an imposed load. The driving is done, using a drop-hammer operated by steam or compressed air in an adjustable guide track, until the pile will no longer drive. Piles are most often installed in clusters of 3 to 10 units and are cut off and capped with concrete pads at elevations desired. These clusters are then joined by steel or concrete "grade beams" to provide a continuous wall foundation. Pile driving may cost from $2.50 to $6.00 per lineal foot, depending on the type of pile used, and is not normal for residential work.

Cassions are another type of foundation footing that may only be used occasionally in residential work. Essentially, cassions are round holes bored into the earth with a machine resembling well-digging equipment, filled with concrete, and bridged between with grade beams to form a foundation system. Cassions are bored to solid natural bearing through fill or even to bedrock, and may be "belled out" at the bottom to provide more actual bearing surface. The diameter of holes varies from 24″ to 84″, with 30″ and 36″ most common. The capacity of machines producing a 36″ hole is about 125 lin ft per hour in normal soil at a cost of approximately $30 per lineal foot, including equipment and labor.

Where some poor foundation conditions due to expanding clay (adobe) soil have been found, a number of contractors have used precast post-tensioned concrete slabs with minimum edge footings. The slabs are poured with plastic tubes installed at about 36″ to 42″ on center (o.c.) in each direction to provide a passageway for wire post-tensioning. After the slabs have set, the wire is installed, secured at one end, and jacked to provide a tensioned reinforcement in each direction. Advocates of this system maintain that if earth swelling or shrinking occurs, the entire slab will act as a unit, thereby minimizing floor cracking due to uneven movement. Cost varies widely but is approximately $2 per square foot of floor area, and may be somewhat offset by the reduction in cost of digging exterior wall footings and filling them with concrete.

## 3.7 PAVING

Another major item of site work is the paving of drives or parking lots with asphalt or concrete and the installation of curbs, gutters, and sidewalks. Most states and cities have standards regarding size, shape, and other details regarding this work and quite often have a list of installers approved by these agencies to do this type of work. In any case, paving work requires expensive equipment, good experience, and is usually subcontracted. Base courses of crushed rock or gravel, 4" to 6" thick, are most often included as a part of paving work, as is the fine grading to exact elevations required. Rough grading to a tolerance of 0.10 ft is considered as grading work or excavation.

Bituminous paving (asphalt or "black-top") material is a hot mix of liquid asphalt combined with graded sand and gravel, and is installed and finished while hot with hand tools and heavy rolling equipment. Its weight is normally about 145 lb per cubic foot, so it is not a lightweight material. This type of paving is installed over a base course of crushed rock or pit-run gravel rolled to required elevations and may be from 1" to 3" thick, depending on the surface use. Both base course and topping are purchased in tons, but it is more convenient to calculate requirements in surface square feet.

Concrete paving, sidewalks, and curbs are usually sublet to contractors in the paving business. Unless the earth is particularly poor, concrete paving is usually placed directly on graded and compacted earth, and may or may not be reinforced with rods or mesh. Forms are required for sidewalks and curbs but precast or preformed gutters may mark the extremities of concrete paving. Some eastern cities use precut granite blocks for curbs, which will cost approximately 50% more in place than concrete. Concrete sidewalks are formed with 2 × 4's as edges, or "bender boards" on curves, so

**Table 3-2**  Estimating paving

| Operation | Quantity | Sq Ft/ 8 hr | Average Cost (dollars) |
|---|---|---|---|
| Base course, rock or gravel | | | |
| 4-6" | Sq yd | 3000–4000 | 0.85–1.50 |
| 1" thick bituminous paving | Sq yd | 3000 | 2.00–3.00 |
| 6" thick concrete, no | | | |
| curbs/gutter | Sq yd | 1800 | 10.00 |
| Concrete curb/gutter, | | | |
| 18" wide × 6" high | Lin ft | 100 | 5.50 |
| Concrete sidewalk, 4" thick | Sq ft | 600 | 1.50 |
| Brick on sand bed | Sq ft | 80–100 | 3.00–4.50 |
| Flagstone, random sizes | Sq ft | 80–100 | 3.00–4.00 |

are nominally 4″ thick and may have a number of different finishes, troweled or broom finish being most common. Table 3-2 lists the average cost of various types of paving.

## 3.8 FENCES

Temporary fences required during construction are considered a part of project expense and are often made of stucco mesh (chicken wire) or of plywood. On residential or small projects such fences are a very minor cost, if used at all, but on major projects may involve covered walks, lighting, gates, and even fire protection. Chain-link fences of #9 gauge wire, using steel pipe or H-section posts spaced at 10′ o.c. and 6′0″ high, is the most common type of metal fence for permanent installation. Corner posts, gates, and extra height will add slightly. Metal fences may also be had with vinyl-covered wire, metal pickets, interlaced metal slats, and as security fences with barbed-wire tops. Wood fences may be most any design or height, of various materials, prefabricated or job installed. Output per day for two men is about 150 lin ft average.

## 3.9 LANDSCAPE WORK

Another portion of site work is landscaping, which may include the installation of irrigation systems, ground-cover planting, specimen trees or shrubs, construction of pools or fountains, or no work at all. Most present-day speculative housing goes either with "complete beautiful landscaping as advertised" or bare unprepared earth which the new owner may landscape as he desires. The term "complete beautiful landscaping" may be descriptive of a really good job with growing lawn (sod) or other ground cover, a nice arrangement of shrubs, and one or more half-grown trees, or may be unfertilized earth with grass seed barely germinated and showing a little green plus a few plants in 1-gallon containers. Landscape work on major projects is usually designed by a landscape architect but on many housing projects is a dumping ground for surplus or weak plants from the lowest-bidding nursery.

A good landscape job requires complete preparation of the existing soil or replacement of it with good imported top soil. The soil must be cultivated, fertilized, fine graded, and perhaps mulched to prepare a proper growing base for most ground cover. The contractor should supply material and properly install topsoil, fertilize, and seed grass in a 6″ deep cover and properly maintain for a period of at least three months. If more immediate "show" is desired, many developers use farm-grown sod, strips of growing grass with earth and roots intact, 12″ wide × 2″ deep, which are placed over graded natural earth.

Bushes and small plants are usually specified and supplied in 1- or 5-gal cans and vary by species from 8″ to 24″ high. Trees and large bushes may be supplied in wooden boxes or as "balled-and-burlapped," which means that plant roots are in an earth ball with a burlap covering protecting them. Landscape plant costs vary widely depending on species, age, size, and availability, but small 1-gal sizes of common plants may be had for as little as $2 each in place, while a 15-ft-tall tree may cost $200 or more.

**3.10 OTHER SITE WORK**

Site work on commercial or industrial projects may also include subsurface drainage using large pipe or tunnels, railroad track, wharf or dock facilities, and other special construction. For the average residence there may be some retaining walls done as a part of concrete or masonry, utility trenches or septic system done as part of mechanical/electrical work, masonry or concrete planters, exterior wood decks, swimming pools, or other work done by trades normally doing that type of construction. Most of these items are done by specialty subcontractors who are experienced and have lump-sum prices for their work.

## PROBLEMS

**3-1.** Make a take-off, by the grid system, of the residential site plan drawing on page 194 to provide a no-excess, no-import, level building site.

**3-2.** Calculate the cost of driveway and sidewalks in place.

**3-3.** Estimate the cost of grass cover if seeded; the difference if sodded.

# chapter

# 4

# Concrete and Masonry

Concrete and masonry work are kindred in that they both provide fire-resistant construction, require a mixture of cement and sand or rock aggregate, provide a good sound barrier, and are quite permanent. They are different in appearance, installation, use, and cost. Concrete is a homogeneous mixture of cement, aggregate, and water placed in forms, perhaps reinforced with metal rods or mesh, and finished in a number of different manners. Masonry is a process of using preformed units of stone, clay, or concrete, securing them together with cement mortar or grout, and possibly reinforcing the construction with rods or cross-ties at joints if required by codes. Brick or stone work is usually not further finished with other materials, but concrete units may be plastered over, painted, or otherwise surface finished. Brick and stone may be used as a veneer over other structural walls, but concrete units are seldom used for this purpose. Concrete is usually calculated, provided for, and paid for in quantities of cubic yards. However, masonry is usually calculated in face square feet, but the number of units and amount of mortar or grout must be figured by the piece or cubic yard. Concrete work is often done by the general contractor, but masonry work is essentially done by a subcontractor.

## 4.1 CONCRETE FORMWORK

Almost without exception concrete is placed in forms of one type or another to determine its final shape. The possible exception to this form requirement is where earth excavations may be used as forms or where existing walls or other surfaces provide perimeters. Forms for concrete may be of boards, plywood, metal, plastic, or any other combination that will contain the heavy plastic mix until it achieves enough strength to retain the shape desired. Slabs on grade need only edge forms, but concrete floor slabs that are not otherwise supported need solidly built forms to determine their shape and location.

## 4.2 SLABS ON GRADE

A great many buildings are now being constructed without basements or deep extensive foundation systems. This means that foundations are generally "grade beams," continuous concrete footings which are slightly wider and without vertical stems, partially formed by the earth excavation and partially formed above the adjacent outside earth level. Figure 4-1 indicates a typical exterior footing and an interior footing beneath a bearing wall.

Forms for this type of foundation are very simple; none are required for the portions below grade, as the concrete is placed directly against the preexcavated earth. The portion above grade is formed with 2″ thick lumber of the height required, staked in place, and possibly braced if more than 8″ high. Lumber may be ordered "random length," which means that each piece may be anywhere from 6′-0″ to 16′-0″ long or longer. Plywood would not be economical, as cutting of the large 4′ × 8′ sheets is expensive in labor costs, usually is not reusable in the long narrow strips that result, and needs more solid backing and bracing unless ¾″ thick or more. Some tract developers, who use typical lengths repeatedly, may find it economical to use prefabricated wood or metal forms for slab edges. Estimating lumber for this type of forms involves simply calculating the amount of board feet required to contain the concrete within the perimeters of the plan.

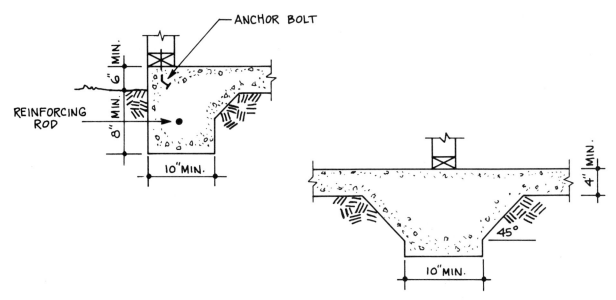

**Figure 4-1**   Grade beams.

Labor for this type of form is relatively inexpensive. Cutting of lumber is minimal, as corners may be formed by simply overlapping or extending one board past the other. Placement labor and staking of the forms should be estimated at 3½ hr per 100 lin ft, with another 3½ hr to fabricate, clean, and move. Forms for this type of work may easily be reused four or five times, and may even be extended to as many as 10 times if stripped and cleaned carefully.

## 4.3 FREE-STANDING WALLS

Concrete walls above the earth grade or for basements will require forms for both sides of the wall to hold the concrete in the proper position. These forms are usually faced with boards or plywood against which the concrete will be placed. The form facing must be reinforced with vertical studs much as a normal wall is constructed, and with horizontal "walers," usually doubled 2 × 4's at about 4'-0" o.c. In addition, these walls need bracing to prevent them from falling away from each other and ties or stretchers to keep them the proper distance apart. See Figure 4–2 for a typical concrete wall form. Similar forms are necessary for basement walls below grade. In place of the details shown in Figure 4–2, some contractors use prefabricated form units 2'-0" wide by various heights, with plywood faces and steel edge and back reinforcing, which bolt together.

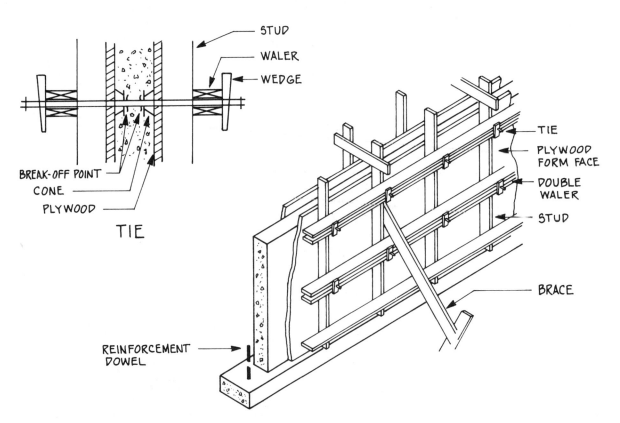

**Figure 4-2**   Typical concrete wall form.

Estimating the amount of "contact surface" (the portion in actual contact with the concrete) is simply multiplication of the length of the wall by the height of the wall on *both* faces of the wall, inside and outside. Consideration must be given when plywood is used as the material comes to the job site in definite-size sheets. If boards are used, approximately 15% waste to cover odd lengths, cutting, and the difference between nominal and actual size must be allowed. Studs and walers, generally 2 × 4's, may be estimated as random lengths taken from the actual lengths required. Form faces now will be held apart as well as together with metal form ties which may be later removed or broken off to allow a portion to remain buried in the concrete. Wood "spreaders" together with wire ties are not often used on major construction. Ties are customarily spaced at about 4'-0" o.c. on walls and much closer for columns or odd shapes, and are estimated at about $100 per thousand (M), including ties, cones, and wedges. Wedges, at $70 per thousand (M), are about 90% reuseable, but ties and cones are lost in the construction. Despite the fact that many parts of formwork may be estimated separately, many estimators rely on the amount of contact feet to determine costs of forms. Wall forms with plywood facings can be calculated at a daily output of 250 to 350 sq ft using a crew of three carpenters and one laborer. Lumber required for wood forms will average about 2½ to 3 times the square-foot area (board measure) of the contact surfaces and may be reused up to four times if carefully stripped and cleaned between uses.

## 4.4 CONCRETE FLOOR SLABS

Residential construction does not regularly require concrete slabs in a horizontal or floor position except where supported on the earth. Slabs above grade require forms when being installed and may be flat-slab type fabricated from plywood, preformed tees which require no on-site work except placing and a little patching, or in unique cases by use of metal or plastic pans. All of these methods are extensively used in commercial and industrial construction. The floor area may be calculated rather easily, as lengths and widths are known; however, forms of any type must be retained in place with horizontal planks, which in turn are supported by 4 × 4" shores spaced at about 4'-0" each way. All heads of wall openings and horizontal beams are similarly supported until fully cured. Some typical forming is shown in Figure 4–3. All types may be reused at least four times. Metal or plastic pans used for "waffle slab" or joist-type work are usually rented.

## 4.5 REINFORCEMENT

Almost every type of material has been tried as "reinforcement" in connection with concrete. Included in the possible list is bamboo, fibers, cloth, and a number of others, but steel, either as rods or mesh, is most common. The reinforcement not only complements the strength requirements for concrete but also acts as a quick transfer agent for temperature changes. Rods are designated by ⅛"

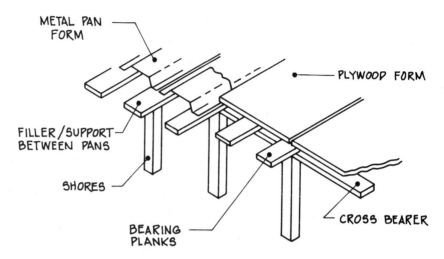

METAL PAN FORM

PLYWOOD FORM

FILLER/SUPPORT BETWEEN PANS

SHORES

BEARING PLANKS

CROSS BEARER

**Figure 4-3** Above-grade slab forms.

increments, now are normally round in shape, and "deformed" with a variety of bumps, ridges, and rings in sizes larger than 3 (⅜″). Sizes, weights, and other details are standardized under ASTM A-615 designations and are indicated in Table 4-1.

"Mesh," or more properly "welded wire fabric," is reinforcement using wires of different sizes to form a grid, with intersections welded. Mesh was formerly indicated by wire size and grid pattern such as 6/6-10/10, which means that the wire formed a 6″ × 6″ grid and #10 wire is used in both directions. The newer designation for the same fabric is 6/6-W.14 × W1.4. Mesh has less value than rods, so in most cases is simply used to distribute temperature changes. Mesh is not normally used for footings or foundations or in vertical walls, but may be used in slabs in addition to rods or by itself. Rod is supplied in standard 20′ or 40′ lengths; mesh is usually supplied in 100-sq yd rolls 6′-0″ wide.

Concrete reinforcement requirements are established by codes and should be indicated properly on the construction drawings. Some builders use one or more rods in footings or mesh in slabs simply to distribute temperature and reduce cracking of the slab even though codes may not require it. Placing and tying of reinforcement is work for an experienced crew whenever it is required by code. Rod is calculated by weight of size required multiplied by length, with an end result in pounds or tons, and may be figured at about $0.25 per foot in place for #4 (½″). Reinforcing rod is lapped at splices by "*X* diameters" of the rod, which means that a requirement of "30 diameters lap" for #4 (½″) is 30 × ½″ = 15″. Lap requirements are vital and vary with the location and use and should be called out. Short dowels and stubs used to connect steel at concrete edges of pours are also important and need to be carefully included in the take-off.

Although anchor bolts and similar small steel or iron items are usually not considered a part of concrete work, these metal items are often figured when calculating concrete. Anchor bolts are embedded

**Table 4–1** ASTM standard reinforcing bars

| Bar size designation | Weight (lb/ft) | Nominal dimensions—round sections | | |
|---|---|---|---|---|
| | | Diameter (in.) | Cross-sectional area (sq. in.) | Perimeter (in.) |
| # 3 | 0.376 | 0.375 | 0.11 | 1.178 |
| # 4 | 0.668 | 0.500 | 0.20 | 1.571 |
| # 5 | 1.043 | 0.625 | 0.31 | 1.963 |
| # 6 | 1.502 | 0.750 | 0.44 | 2.356 |
| # 7 | 2.044 | 0.875 | 0.60 | 2.749 |
| # 8 | 2.670 | 1.000 | 0.79 | 3.142 |
| # 9 | 3.400 | 1.128 | 1.00 | 3.544 |
| #10 | 4.303 | 1.270 | 1.27 | 3.990 |
| #11 | 5.313 | 1.410 | 1.56 | 4.430 |
| #14a | 7.65 | 1.693 | 2.25 | 5.32 |
| #18a | 13.60 | 2.257 | 4.00 | 7.09 |

aSizes 14 and 18 are large bars generally not carried in regular stock. These sizes are available only by arrangement with your supplier.

*Source:* Concrete Reinforcing Steel Institute.

in concrete foundations and secure the wood structure to the base. Bolts are nominally ½″ diameter by 8″ to 12″ long, spaced at 4′-0″ to 6′-0″ around the perimeter of the building. Interior walls may be anchored similarly, but most often powder-actuated guns are used to drive pins through the sill members and into the concrete floor slab.

## 4.6 CONCRETE

There is hardly a building that does not use concrete in some form or other in its construction, and this material is one of the oldest used by man. Basically, concrete is a mixture of cement, sand, heavy aggregate (crushed stone or gravel), and water that is proportioned to provide a stone-like product with a definite compressive strength at an assumed totally cured period of 28 days. The proper relationship of dry materials and the correct amount of water added will give compressive strength from as low as 1000 pounds per square inch (p.s.i.) to 60,000 p.s.i. or more. Most concrete used in residential work and small projects has rather low strength, from perhaps 2000 to 3500 p.s.i. in compression, and no practical value at all in tension. Proportioning of mixes is established by standards of ASTM and ACI (American Concrete Institute) and is available in many books. Very little concrete is now job-site mixed as transit-mixed concrete is less bother to handle and the transit-mix plant accurately measures the ingredients to provide the required strength. Normally, concrete is measured in cubic yards and weighs approximately 150 lb per cubic foot. "Lightweight" concrete, generally used for nonstructural fill, roof decks, and the like, is made with expanded shale aggregate and weighs only about 90 to 110 lb per cubic foot.

Residential work uses concrete for footings for basement walls, the walls themselves, as floor slabs, garage floors, driveways, walks, fences, and similar uses. The concrete may be one strength or different strengths and may have no additional finish after being placed or may be troweled smooth and treated with accelerators, retarders, acid color, exposed aggregate finish, or patterns pressed into the wet surface to represent brick or stone. Walls may be formed to present a board-type appearance, rough vertical columns, brick, or other form-liner impressions, or may be acid treated or sandblasted. Concrete will flow into almost any shape that can be built and will retain that shape indefinitely. Color pigment may be added but is usually only added to material that will be used as a "topping," as color throughout a concrete section is normally not required and is expensive to provide except as a finish coating.

Mixing, placing, and curing of concrete are the three major steps that need to be done properly for a good job. As already noted, job-site mixing is now nearly extinct except on super-large industrial projects. Transit-mixed concrete is available in mobile-mixer trucks with capacities from 3 to 15 cu yd and may be introduced into the forms by direct chute from the mixer, by pumping, by wheelbarrow, or by buggies. In residential/apartment construction most concrete comes directly from the mixer into the forms except that which is finally located above the first-floor level. On multistoried projects, concrete may be pumped, handled in drop-bottom buckets by crane, wheeled in buggies, or transmitted by conveyor. Tilt-up methods, using wall sections that have been cast in place on the ground and then raised into position and joined with cast-in-place columns, are not practical for small jobs. Precast tees, slabs, U sections, or similar parts are not practical either, as they need heavy support and residential work is normally wood supporting walls and columns and not adequate to support heavy concrete loads.

The cost of concrete at transit-mix plants has changed materially in the past few years. Today concrete is not the inexpensive material it was once considered to be. Moreover, quantity discounts are often given only to users of concrete on big projects. In 1977, transit-mix concrete averaged $13 per cubic yard at the plant. In 1979 this cost had jumped to nearly $30, and now ranges from $50 to $75 per cubic yard. Relatively higher costs are necessary when concrete is to be placed above first-floor levels. Finish of concrete floor slabs by troweling costs about $0.25 per square foot, while finish for exposed aggregate runs about $0.25 to $0.45 per square foot. Floor color, dusted on wet surface and troweled in, averages about $0.40 per square foot for most colors.

Curing of concrete is as important as mixing and placing. Retaining the moisture within the concrete during the curing period is accomplished by spraying exposed surfaces with various types of material, principally petroleum derivatives, or by covering with burlap, sand, or straw. On some large slabs irrigation sprinklers may be used to keep the concrete damp. Cost for curing varies from $0.05 to $0.30 per square foot. Also considered as concrete work is the installation of waterproof membrane beneath slabs. This may be 15-lb black saturated felt, 8-lb red rosin paper, or more usual at present is 0.006-mil Visqueen plastic, and costs about $0.07 to $0.10 per square foot and requires 10% "waste" for laps and edges. Membrane is normally installed over a leveled sand base about 2″ to 4″ thick, also considered part of concrete work in many cases. Insulation enclosed by concrete in footings and slabs is not considered concrete work; it is discussed in Chapter 7.

**4.7 FORM REMOVAL**   In most instances concrete is retained in the forms until final set and partial curing have occurred. In cases of slab-on-grade this may be as little as 3 days or in cases of concrete beams may be as long as 28 days. Of primary importance is the reuse of forms or dismantling as waste or for use in other locations of the construction. In some areas stripping of forms is a labor union problem; forms to be reused require carpentry labor, which is the normal situation, but if forms are to be dismantled or wasted it requires common labor.

**4.8 MASONRY**   Masonry includes stone work, brick, and concrete units, used primarily as foundations, exterior or fire-separation walls, or as veneer over other construction. Brick and concrete units are manufactured in standard sizes as shown in Figure 4–4, while stone may be any size, thickness, quality, or color. Masonry is installed with cement mortar at bed and end joints, usually ⅜″ or ½″ thick. Brick walls are laid up with a space between wythes (separate vertical parallel walls) and connected with occasional cross bricks or metal ties. This method provides "cavity walls." Concrete units are laid in a similar manner, but obviously there is no open space between inner and outer shells; each unit has an open core. In areas of possible earthquake damage the "cavity" in brick work and the open cells in concrete units are reinforced with standard reinforcing rods and fully grouted with a soupy mixture of concrete. Normal spacing for vertical reinforcement is #4 at 24″ o.c. with #4 at 48″ horizontal fully encased in grout up to 10′ high; reinforcement requirements should be shown on the drawings for other situations.

**4.9 BRICK WORK**   Clay bricks are available in a variety of sizes and colors, mostly red to brown shades, or as glazed brick with one face glazed in color. Brick are used for walls, fireplace construction, paving, and as a veneer. In addition to the actual brick units involved there are many accessories, such as clay flue linings, fireplace dampers and ash dumps, fire brick linings, masonry reinforcement, and various lintels and ties. Brick is laid in various face patterns shown in Figure 4–5, which affects the cost of installation. Costs for masonry construction depend a great deal on location of the masonry working deck as well as on the availability of labor and material. Starting at ground level a masonry wall may be laid as high as 4′-0″ with reasonable accessibility; however, scaffolding at intervals of about 4′-0″ is necessary above that to install work properly at higher levels.

Clay bricks used for floors or patio paving may be laid in

DIMENSIONS AS SHOWN ARE ACTUAL UNIT SIZES.  A 7⅝" x 7⅝" x 15⅝" UNIT IS COMMONLY KNOWN AS AN 8" x 8" x 16" CONCRETE BLOCK.
HALF LENGTH UNITS ARE USUALLY AVAILABLE FOR MOST OF THE UNITS SHOWN BELOW.  SEE CONCRETE PRODUCTS MANUFACTURER FOR SHAPES AND SIZES OF UNITS LOCALLY AVAILABLE.

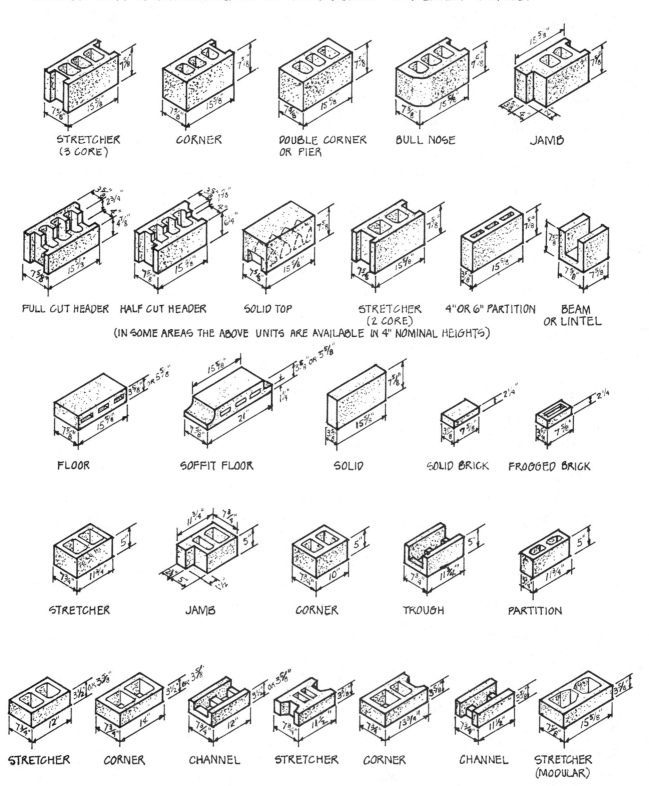

# CONCRETE MASONRY UNITS

**Figure 4-4**  Standard concrete and masonry units.

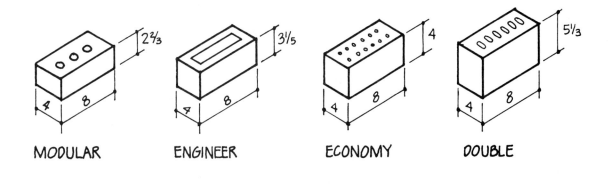

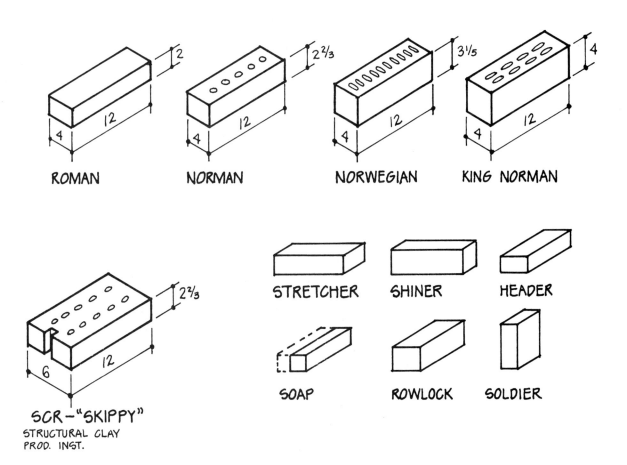

BRICK ARE CLASSIFIED AS **SW** WHERE HIGH RESISTANCE TO WEATHERING IS REQUIRED, **MW** WHERE MODERATE WEATHERING IS EXPECTED, AND **NW** FOR BACKING OR INTERIOR USE. FACE OF BRICK MAY BE SMOOTH, WIRECUT, STIPPLED, RUFFLED, SANDMOLD, OR MANY OTHER TEXTURES.

# BRICK MASONRY UNITS

**Figure 4-4** (Continued.)

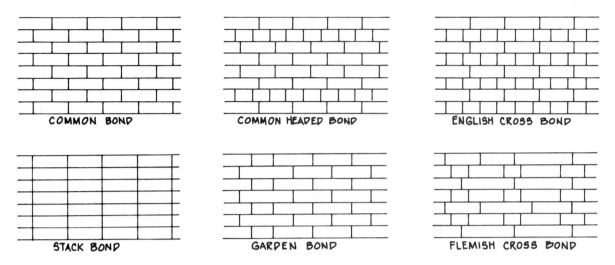

COMMON BOND          COMMON HEADED BOND          ENGLISH CROSS BOND

STACK BOND          GARDEN BOND          FLEMISH CROSS BOND

**Figure 4-5**  Brick face patterns.

cement mortar bed or on a sand bed in a wide variety of patterns. Although there are again a wide variety of flooring types, including acid-resisting, heavy-duty industrial, pavers, and cast ceramics, most residential work uses common brick either flat or on edge, depending on the pattern. Labor may be bricklayers or tile setters, depending on the local union rules, and a journeyman plus a tender should install about 100 sq ft per day.

## 4.10 CONCRETE MASONRY UNITS

Concrete masonry units (blocks) are used in many places where brick or cast-in-place concrete may be used. In most cases walls are formed by one wythe or layer of units of proper size, 8″ or 12″, with cement mortar joints ⅜″ or ½″, similar to brick work. Also in common with brick construction, the cells of concrete units may be left open or may be fully grouted. Where grouting is required to provide a fireresistant wall, all cells are filled, but where earthquake or structural requirements govern, only the cells that contain the vertical reinforcement are filled. A continuous horizontal bond beam, formed with U blocks and reinforced with reinforcing steel, is used at the top of walls, at floor levels, and above openings as lintels. Good design attempts to utilize full blocks, half blocks, special corners, and similar units to reduce excessive cutting. In addition to normal closed-faced blocks of various sizes there are a great many "decorative" blocks with raised designs on one face, perforated 4″-thick units for fences, and many types of split-face and sill units. Most manufacturers produce normal "heavyweight" concrete units as well as "lightweight" units made with expanded shale.

## 4.11 STONE WORK

Stone used in residential construction is almost always as a decorative veneer or at least a very minor item. In some parts of the country stone may be used for foundation walls, but generally concrete units or cast-in-place concrete is cheaper and more readily installed. Limestone may be accurately cut and shaped to provide lintels at openings in brick construction and a limited amount of bluestone (New York–Pennsylvania) and granite (New England–Midwest) may be used. In the southwestern United States cut stone is very limited, but some rubble is used and flagstone may be used for veneer or walks and patio pavings. The pattern in which the stone is laid has the greatest effect on the cost. Figure 4–6 shows some typical stone patterns.

Most stone is calculated and purchased by the ton, although in a few communities it is sold by the cord containing 100 cu ft or by the cubic yard. Rubble stone work requires the least amount of cutting and fitting. A mason should be able to set about 100 to 150 cu ft of rubble per day, using 7 to 9 cu ft of mortar. Random ashlar pattern stone is rough cut, supplied at about 12 cu ft per ton 4″ thick, and provides about 35 to 40 sq ft of wall. Indiana limestone is used for many sills, steps, copings, and so on, and may also be used as a veneer 4″ thick over wood construction. In the southwestern United States a variety of decorative stone, including lavastone, Arizona flagstone, featherrock, and many local stones, is available and is used primarily as veneer.

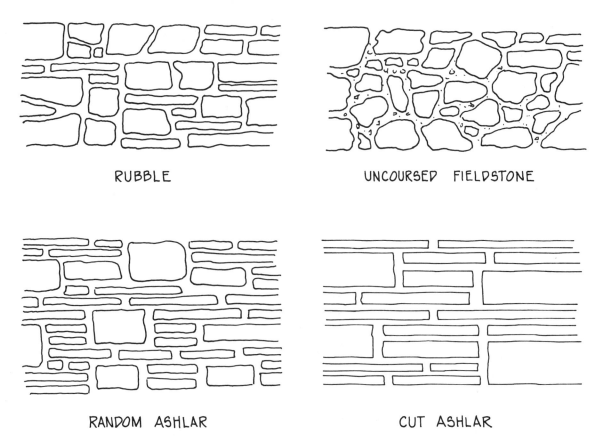

RUBBLE                          UNCOURSED FIELDSTONE

RANDOM ASHLAR                     CUT ASHLAR

**Figure 4-6**   Typical stone patterns.

**4.12 OTHER MASONRY** Brick, concrete units, and stone are not the only masonry that may be encountered. In many locations burned clay building tiles are used as structural material but principally in industrial construction. Some glass block is still used where an obscure sight line is desired with some transmission of light. In churches, banks, libraries, and government buildings a great deal of cut or carved limestone, granite, or other stone may be used. Marble is used as a veneer over concrete or steel construction but seldom appears in residential work. Some slate is used for floor paving, but slate used as a roofing material is not masonry work. To maintain the focus on basic residential estimating, none of the above are included in this book.

## PROBLEMS

**Sample problem:** Estimate the amount of material required for a wall 16'-0" long, 4'-0" high, using 8" × 8" × 16" concrete blocks, reinforced vertically with #4 rods at 24" o.c. and horizontally at top with one #4, installed on a concrete base of 9" × 18" buried to the top of the base and reinforced with one #4 rod horizontally.

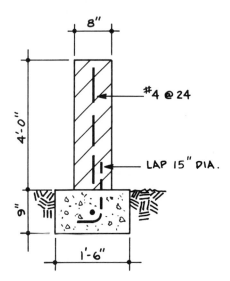

Excavate for footing .75 × 1.5 × 16.0 = 18.0 cu ft

$$\frac{18.0}{27} = .666 \text{ cu yards.} \quad \text{haul total away}$$

Concrete base is identical so use .66 yards concrete

Reinforcement: 9 pc × 48" = 432"

9 dowels 30 diam. × ½" (#4) = 15" + 8" =

23" say 24"  216"

$$\frac{432 + 216 \quad 648"}{12} = 54 \text{ lin ft}$$

54 lin ft × .668 (wt. per ft) = 361 pounds

Concrete blk:  16' × 4' = 64 square feet.    6 courses high
                    ea. 48" requires 3 blk. plus ends
                    12 blk. per course × 6 = 72 blk.
                    Req; 6 half block for ends plus
                          66 full blk.    No waste

**4.1** Estimate quantities for a cast-in-place concrete wall, 10'-0" long, 5'-0" high, 10" wide, reinforced with #5 rods at 24" o.c. each way, installed on a concrete base 12" deep, 20" wide, with the top of the concrete 6" below the finish grade. Required are excavation, back-fill, concrete, and reinforcement.

**4.2** With quantities of concrete from Problem 4-1, estimate the amount of formwork required using ¾" plywood, 2" × 4" studs at 24" o.c., and six 2" × 4" horizontal walers. Forms for a buried foundation are not required. *Note:* Do not forget that forms are required for *each side* of the wall and to close the ends.

**4-3** Using the plans included in this book, what is the in-place cost of the masonry fireplace shown?

# chapter

# 5

# Wood Construction

Wood is one of the most commonly used materials in any construction, especially in residential work. In concrete or masonry structures wood is used for trim, cabinet work, some doors, and miscellaneous decorative work, but in residential building it forms the structural frame, floors, roof construction, exterior covering in some cases, and in most of the doors, windows, and trim. The wood used is from many species of trees from many parts of the world, available as solid lumber, laminated stock, or plywood, and either green, yard dry, or kiln dried. Only a relatively few species are used for construction in the United States; Southern pine (loblolly, longleaf, shortleaf, slash), Douglas fir, spruce, larch, hemlock, western pine (sugar, ponderosa, white), California redwood, some cypress, and various hybrids. Hardwoods such as birch, oak, pecan, and walnut are used for trim, flooring, and veneer work.

## 5.1 MATERIAL

Lumber, solid stock usually smaller than 8″ × 8″, and "timber" sizes, larger than that, is the major wood product used for residential construction as studs, joist, rafters, and other structural members. Producers of lumber have organized into four major grading or control groups: California Redwood Association (CRA), Southern Forest Products Association (SFPA), Western Spruce-Pine-Fir Association (WSPFA), and Western Wood Products Association (WWPA). Mills that produce lumber may belong to one of these organizations and generally produce lumber to standard sizes as recommended under National Grading Rules PS-20-70 and as tabulated in the Appendix. Structural lumber is "yard dry" at 19% moisture content or less, but lumber used as millwork may be kiln dried to about 6 to 10%. Lumber is grade marked by the producer for strength and specified by the designer to require different grades for different uses.

Laminated beams and arches are made from layers of standard lumber glued together in shapes and sizes required. This lamination provides beams that are longer than may be obtained from a single tree, deeper or wider than the tree diameter, and much stronger. Although most laminated timber is in the form of rectangular-section straight beams, it is easily possible to form tapered beams, arch sections, or other shapes. Various appearance grades are obtainable, thereby using almost any type of lumber from clear stock to that with many knots or other defects. Since lamination does not depend on the individual strength of pieces used, the resulting builtup beam is stronger than any possible solid timber unit. Laminated timber is not used widely in residential work, generally being found only in the form of ridge beams for exposed "cathedral" roofs or for "post-and-beam" exposed structural framing.

Plywood also is a process of laminating, but instead of lumber the plys are sheets of wood ranging from ⅟₂₈" to ¼" thick. Due to the process, plywood has odd number of layers, with the outer exposed surfaces having the grain in the same direction and counter to the intermediate "core" sheets. Plywood stock is obtained by rotating a log against a sharp blade, thereby producing a continuous sheet called "rotary" material, or by slicing to produce successive sheets with nearly matching grain. Structural plywood is available in thickness from ¼" to 2", three-ply minimum, is manufactured in grades A, B, C, and D, and is usually from rotary-cut Douglas fir. The American Plywood Association (APA) has established grading rules and mills belonging to this association provide most of the structural plywood used. Sliced stock is usually of a more exotic species, thinner, may be matched in a number of patterns, as shown in Figure 5-1, and is used only for face veneer for interior paneling, furniture, and similar products.

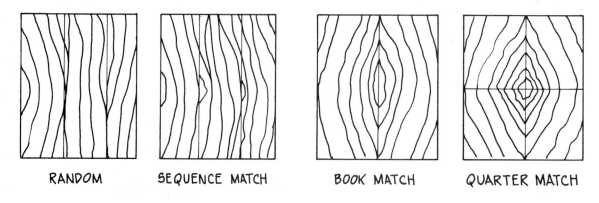

RANDOM          SEQUENCE MATCH          BOOK MATCH          QUARTER MATCH

**Figure 5-1**   Wood veneer match patterns.

## 5.2 RESIDENTIAL FRAMING

Individual preferences or code requirements will dictate the method by which a residence is "framed," that is, the way the structural skeleton is erected and connected. The two principal established classifications are "western" frame (platform frame) and "balloon" frame, as shown in Figure 5–2. In **western** frames a platform is built on the foundation system, the wall studs are set on a bottom plate and capped with a top plate, usually two 2 × 4's lapped at joints, and another floor, ceiling, or roof frame built on the top plates. If studs are erected one at a time they must be supported upright and are toenailed to the bottom plate, but a quicker and easier method is to assemble the wall horizontally and then erect it in place. In this manner the studs are end-nailed through the bottom and top plates, openings are readily framed, and the skeleton is virtually complete when correctly placed and raised into a vertical position. Workmen are active on a solid platform and there is little need for scaffolds, ladders, or extensive bracing. When all wall sections are erected and properly connected, the fire blocking between studs and other miscellaneous work is easily accomplished.

**Balloon** frame employs continuous studs from the mud sill atop the foundation to the top of the building, with floor joist installed on top of a horizontal ribbon and nailed directly to the studs. This type of framing requires long sticks of lumber for studs, much bracing to retain the studs in position, considerable amounts of scaffolding or ladders so that men can work, and much cutting and fitting of floor material and horizontal bearing ribbons. Blocking is installed at floor levels to provide "nailers" for floor covering and to block off any "flues" between the studs. With some luck, studs, ribbons, and cap plates may be assembled horizontally and hoisted erect, but again these framed sections are more than one-story high, loosely connected and braced, and awkward to handle until fully erected and securely braced.

Post-and-beam framing is exactly what it indicates. Vertical columns are spaced at intervals along the walls or beneath ridges, and horizontal beams, often of laminated timber, are secured to the columns to form the building frame. Most often the floor construction is a concrete slab on grade and the vertical posts are secured with a wide variety of U straps, angles, or pins. The horizontal members are in turn secured to the posts with angles, bent plates, or similar connections with bolts or lags. The actual walls, interior as well as exterior, are nonbearing (not supporting loads) and are simply fillers of spaces between columns and may be of stud construction, masonry, glass, or most any material. Most post-and-beam construction is designed to use the structural frame as an exposed design element and is limited to timbers available.

## 5.3 LUMBER TAKE-OFF

Most lumber is cut from green sap-filled logs to approximately nominal sizes, then "surfaced" or planed on faces and edges to actual size. This means that a 2 × 4 is *not* 2″ × 4″ when finished

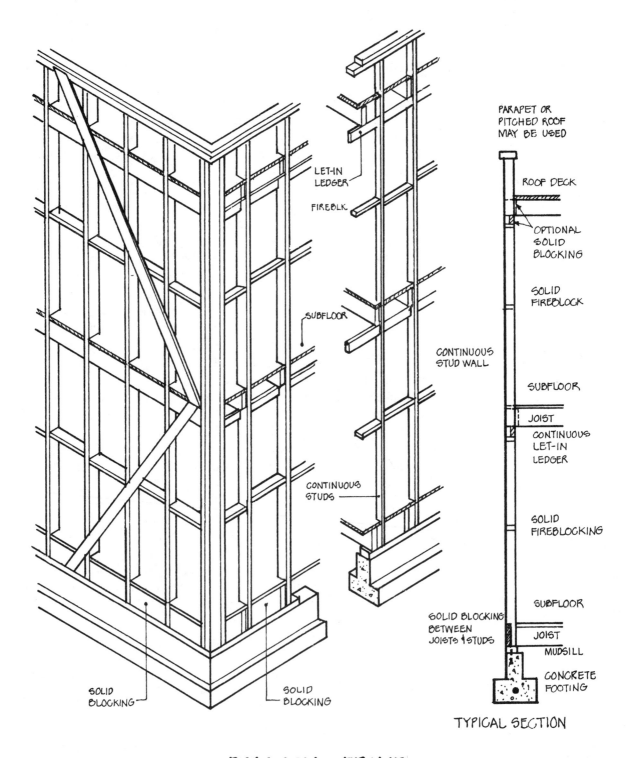

LET-IN
LEDGER

FIREBLK.

SUBFLOOR

CONTINUOUS
STUD WALL

CONTINUOUS
STUDS

SOLID BLOCKING
BETWEEN
JOISTS & STUDS

SOLID
BLOCKING

SOLID
BLOCKING

PARAPET OR
PITCHED ROOF
MAY BE USED

ROOF DECK

OPTIONAL
SOLID
BLOCKING

SOLID
FIREBLOCK

SUBFLOOR

JOIST

CONTINUOUS
LET-IN
LEDGER

SOLID
FIREBLOCKING

SUBFLOOR

JOIST

MUDSILL

CONCRETE
FOOTING

TYPICAL SECTION

# BALLOON FRAME

**Figure 5-2**   Residential framing methods.

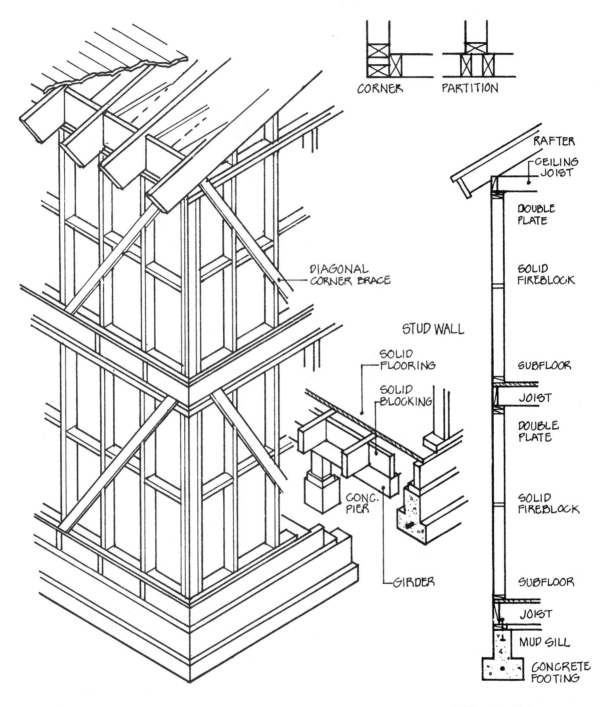

CORNER    PARTITION

DIAGONAL
CORNER BRACE

RAFTER
CEILING
JOIST

DOUBLE
PLATE

SOLID
FIREBLOCK

STUD WALL

SUBFLOOR

SOLID
FLOORING

JOIST

SOLID
BLOCKING

DOUBLE
PLATE

SOLID
FIREBLOCK

CONC.
PIER

SUBFLOOR

JOIST

GIRDER

MUD SILL

CONCRETE
FOOTING

TYPICAL SECTION

# WESTERN FRAME

**Figure 5-2** (Continued.)

LUMBER LIST

| PIECES | MATERIAL | GRADE | SIZE | LENGTH | BM |
|--------|----------|-------|------|--------|------|
| 6 | Redwood | Heart | 2x6 | 10-0 | 60- |
|  |  |  |  |  |  |
| 10 | Pine | #1 | 1x6 | 10-0 | 50- |
|  |  |  |  |  |  |
| 8 | Doug. Fir | Constr. | 1x6 | 10-0 | 40- |
| 50 | " " | " | 2x4 | 8-0 | 333- |

**Figure 5-3**  Lumber list.

two edges and two sides (S4S—surfaced four sides) but is 1½″ × 3½″. Other sizes are shown nominal as well as actual in tables in the Appendix. Lumber is calculated, purchased, and used as *board measure* (BM or BF) with each *board foot* equivalent to a board 12″ × 12″ × 1″ thick. This means that a 1″ (nominal)-thick board 12″ wide (nominal) by 10′ long is figured as 10 BM even though the actual board is ¾″ thick and 11½″ wide. By the same token a 2 × 6 (nominal) 10′ long is also 10 BM, as it is equal to two 1″ x 6″ boards laid edge to edge. Plywood is *not* figured in board feet but in surface square feet and is priced in accordance with the thickness. Grading of lumber and plywood is concerned primarily with the number of defects, such as knots, splits, checks, and warp and other disfigurements. Each producer association has printed grading rules that may be obtained free or at minimum expense. Horizontal framing members, such as joists and beams, are usually one or two grades better than vertical studs or columns since the structural value of a wood member is less with the load applied perpendicularly to the wood grain. In lumber take-off and ordering, each size, grade, and species is kept separate and calculated in board measure, as indicated in the partial list shown in Figure 5–3.

## 5.4 FLOOR CONSTRUCTION

Floors are horizontal diaphragms made by installing boards or plywood over floor joists. In addition to providing a working platform during construction and a final walking surface, the resulting horizontal plane helps keep the building in shape in case of settlement, wind, or earthquake. Floor joists are used to span from the foundation walls to intermediate supports, are usually spaced at 16″ o.c., and are blocked or X-braced between supports for additional rigidity. Sheathing over a joist may be 1″ nominal thickness boards, usually installed diagonally at 45° for resistance to racking, or solid plywood. Where plastic tile or sheet material is to be installed in baths or kitchens, a smooth-surfaced compressed board is used as an overlay to provide a better surface.

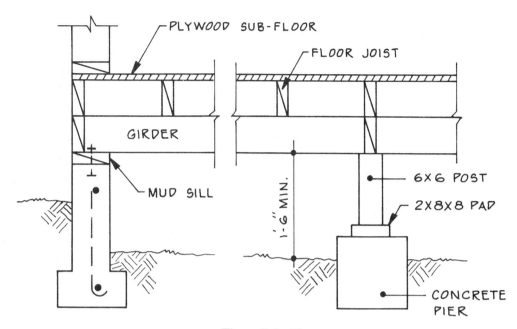

**Figure 5-4** Floor supports.

Material estimates for floor construction may be divided into two major items: floor supports and floor sheathing. Floor supports start with the wood members installed on the foundation beneath the joists and as girders between foundation walls (Figure 5-4). Mud sills installed on top of the foundation walls are usually random length of sizes required or may be standard lengths that fit the foundations. Be sure to list mud sills and other similar parts that are pressure-treated to reduce decay or are of a special species (California redwood or cypress). Supporting joists should be considered for lengths that will economically span between supports and reduce amount of cutting or waste. Blocking or rim joists (the joists or members across the open end of joist installations) may be random lengths or of a lesser quality but often ordered the same as joists to ensure that any lumber intended for use as joists but which may have large knots or other defects may be replaced by good stock (the blemished lumber is used for blocking). Joists are usually spaced at 16″ o.c., so each 4 ft of spacing will include three joists, plus one end joist at the end of the run. Thus a span 20′-0″ long would require 5 × 4′-0″ = 15 joists *plus* one end joist, or 16 total joists. This method may be used for ceiling joists or any other members spaced at 16″ o.c.—but **don't forget the final end member.**

## 5.5 FLOOR SUPPORTS

Members supporting the floor covering (sheathing or subfloor) are normally installed on a mud sill placed on top of the concrete foundation stem wall or concrete-block wall. When a concrete slab is used for the floor surface, obviously no supporting members or wood floor sheathing is required. The mud sill is usually pressure treated for decay resistance or is "foundation grade" California redwood and is bolted to the concrete with ½" diameter bolt at 48" to 60" o.c. Plastic film, asphalt-treated paper, or caulking material is applied on top of the concrete to separate the wood from possible moisture in the concrete. This item is often overlooked but should be estimated by lineal feet of concrete wall at about $0.50 per lineal foot in place. Bolts will cost about $0.50 each with another $0.50 for installation.

On top of the mud sill and perpendicular to the wall the floor joist are installed to support the floor sheathing or subfloor. The inner ends of the joist, when they do not extend from wall to wall, are supported on girders, usually 4" × 6" or 4" × 8", set on concrete piers. The exposed ends are connected by blocking installed solidly between floor joists or by a continuous "rim joist" across the face of the joist. The floor system is further braced with solid blocking, wooden cross bracing, or steel cross bracing at approximately 8 ft along the joist length.

Labor for floor framing will be carpenters, with perhaps one laborer to help supply materials. In larger construction much of the lumber cutting will be done by a specialist using radial-arm saws at a permanent location. On smaller projects most cutting will be done with portable electric-powered "skill saws." A crew of two carpenters should be able to install about 250 to 300 lin ft of 2" × 6" mud sill per day and about 100 lin ft of 4" × 6" girder in 3 hr or about 250 lin ft per day. Floor joists are normally 2 × 6's spaced at 16" o.c. and should be calculated at about 3 hr for 100 sq ft of floor area. Solid blocking between joist requires about the same amount of time, 3 hr per 100 lin ft of blocking. If wooden X-bracing is used, it requires about 4 lin ft of 1" × 3" material per set and about 10 minutes per set to install. Common 16d wire nails are used for most floor framing and about 350 nails or 7½ lb are required for 1000 sq ft of floor framing area.

## 5.6 FLOOR SHEATHING

Today most floor diaphragm or decks are of plywood supplied in 4' × 8" sheets so will have to be calculated to the nearest full sheet. The best method to figure plywood would consider installation of the long 8' edge parallel to the supporting joist, as shown in Fig. 5-5. The major problem with installation in this manner is the fact that the plywood supplies a less springy deck when installed with the surface grain perpendicular to the joist (Fig. 5-5). Where plywood ⅝" or thinner is used for decking it is wise and customary to provide solid blocking along the edges of plywood that is not directly over a supporting member. The cost of labor and materials for installing such blocking should be carefully considered, as it may be more economical to eliminate the blocking by the installation of thicker plywood.

When floor sheathing or subflooring is not of plywood, 1" nominally thick boards 6" to 8" wide are used and may be installed diagonally at 45° across the floor joist or at 90° to the joist. Diagonal installation gives more resistance to racking or "out-of-

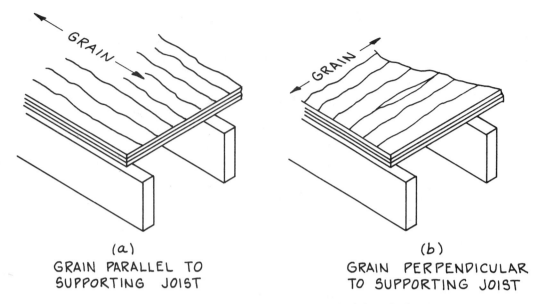

(a)
GRAIN PARALLEL TO
SUPPORTING JOIST

(b)
GRAIN PERPENDICULAR
TO SUPPORTING JOIST

**Figure 5-5**  Plywood deck installation.

square'' distortion due to settlement or similar causes. Cut-off ends and waste for board sheathing should be figured at about 20% more than the floor area.

Plywood floors are installed more rapidly than board sub-floors, due primarily to the larger area covered by each unit. This means that approximately 7 to 8 hr of carpenter labor per 1000 sq ft of area is needed for plywood. In contrast, only about 750 to 800 sq ft of boards laid diagonally can be installed in 8 hr. Underlayment, required by FHA Minimum Property Standards when the finish floor is of resilient flooring, carpet, or ceramic applied with adhesives, is usually ¼" thick hardboard such as Masonite or ⅜" thick particle board and is installed at about 1500 sq ft daily output. Nails are usually 8d common wire nails for sheathing and 2500 to 3000 (17.5 lb) are needed for 1000 sq ft of area.

**5.7 WALL CONSTRUCTION**

Most walls in residential work, and in many other small projects, are framed with wood studs placed at 16″ o.c. with a bottom continuous plate and a double top plate. Unless there is some special requirement for additional strength or room for pipes or ducts, this frame is composed of 2″ × 4″ nominal size lumber. In addition to the bottom plate and the two top plates, there is usually a row of blocking, often called "*fire blocking*," at 4-ft intervals of height or at approximately midpoint in normal construction providing 8 ft ceiling height. Double or triple studs are required at openings, wall intersections, and at wall corners. Stud requirements are often estimated at one stud per lineal foot of wall without openings and plates at three times the length of the walls. Unless there are complicated wall arrangements, this estimate will supply three studs at corners and studs and cripples at openings.

Studs are usually ordered either a 96″ nominal length or may be ordered precut to provide 8-ft ceiling clearance when plates are installed. Top and bottom plates may be figured random length. In most cases this will work out satisfactorily, but remember that top plates must overlap at least 48″ at each joint along a wall in order to give desired rigidity to the wall. This may mean that the lumber list may indicate specific lengths even though the total length and Board Measure amount is unchanged. Headers or lintels over door or window openings are usually of the same thickness as wall studs but may be 6″ to 12″ deep, depending on the opening width and should be taken off separately from studs and plates. The required sizes should be marked at all openings or by general notes on the drawings. Normal stud wall construction will require about 110 Board Measure for each 12′-6″ length of 2″ × 4″ wall 8′ high and about 165 Board Measure for the same length and height if 2″ × 6″ studs are required.

Carpentry labor will average about 2½ to 3 hr for each 100 sq ft of wall with additional labor of 12 hr for 100 lin ft of 4″ × 6″ headers or 9½ hr if headers are 4″ × 12″. Another method used to figure labor is in relation to thousands of board measure (BM) feet of lumber required. By this method about 1½ to 2 hr is required to place 1000 BM (MBM) of studs and plates, an additional ¾ to 1 hr for headers over openings, and another ½ to ¾ hr for fire blocking per MBM. Bracing, included in the 100-sq ft method, requires about 1½ to 2 hr installed "cut in" at studs for each 100 lin ft. Approximately 500 common 16d (10 lb) of nails are required per MBM of lumber.

## 5.8 CEILING CONSTRUCTION

Ceiling construction is much like floor construction and may in fact be floor construction in multistory projects. The primary difference is in the size of joist used, as ceiling joist must span from wall support to wall support. The sizes of joist and their direction of run should be clearly marked on the drawings. In multistory construction some special framing may be necessary at stair openings or similar locations. Joist are installed on the top of wall plates, blocked at open ends, and braced between if the length is greater than 8′ between supports.

Joist will usually be larger than 2″ × 6″ and may be up to 2″ × 14″ if required by the span. Joist 2″ × 6″ spaced at 16″ o.c. require about 2½ hr of carpenter labor per 100 sq ft of floor or about 70 minutes per MBM. If joist are 2″ × 10″ at 16″ o.c. labor should be about 3½ hr per 100 sq ft of floor or 1½ hr per MBM. Usually, no allowance is made for openings smaller than about 60 sq ft and cross bridging or blocking between joist should be figured in the same way as indicated in section 5.4. It should be obvious that in multistory construction the first-floor ceiling requirements are calculated in a similar manner, but the second-floor subfloor needs to be figured in a manner similar to first-floor construction. Nail requirements are the same as those needed for first-floor joist supports and subfloor.

**5.9 ROOF CONSTRUCTION**

Roof construction can be flat, gabled, hip roof, gambrel, or any of a dozen or more standard variations. The basic names of roof members is shown in Figure 5-6 for an L-shaped house with two intersecting roofs. Intersecting roofs should both be at the same pitch for easy framing, but sometimes this is not possible, so watch closely for the roof pitch symbol $\overline{\phantom{12}}^{12}\rceil_4$ , in which the 4 indicates the number of inches of rise for each 12″ of horizontal run. For flat or simple roofs the length of the rafter must be known and should be taken off by measuring along the line of the rafter or by mathematical calculation. Once this length is known, multiply by the length of the roof and you will have the number of square feet of roof area. If the roof is a gable, be sure to multiply this answer by 2, as there are usually two similar sides to a complete roof. For the area of intersecting roof, make two or more calculations of the roof planes and add them together. You should now have the length of the rafters required as well as the total area for roof sheathing. Rafters are usually spaced at 24″ o.c., so divide the length into 24″ increments and add one for the open end. Be sure to allow for the wall overhang required. Save the total area figure, as this should also be the amount of roofing material needed to cover the roof.

Roof rafters, hips, and valleys are usually 2″ × 6″ or 2″ × 8″ for most residences. There is also a longitudinal ridge member that is customarily 2 in. deeper than rafters and there may be rafter ties spaced about 48″ o.c. to connect the opposite rafters together horizontally above the plate line and below the ridge. Additional 1″ × 4″ or 2″ × 4″ stock may be used as bracing from plates of interior walls to longitudinal supports, along the bottom of the rafters, or con-

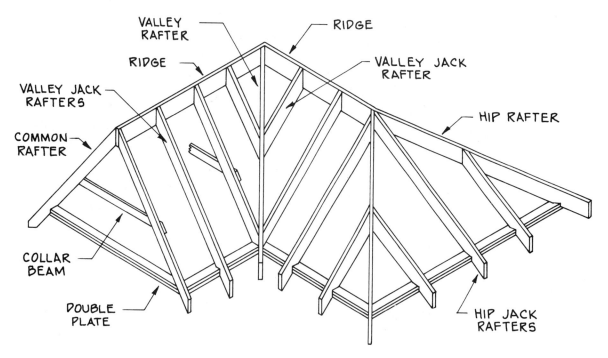

**Figure 5-6**   Typical roof framing.

nected directly to the rafters. These ties may be eliminated if the ceiling joists are installed parallel with the rafters and connected to them at the plate line.

Roof sheathing may be solid covering of edge-matched boards, of plywood, or of spaced sheathing. The first two provide the solid surface required for all types of composition roofing, while the spaced sheathing is often used when shakes or wood shingles will be the roofing material. In the latter case 1″ × 4″ boards installed parallel to the ridge and eave lines with a 4″ space between boards is used and often indicated on the drawings as "1″ × 4″ sheathing spaced at 8″ o.c." When eave overhangs are exposed, all decking on the exposed portions is installed solid for appearance. If boxed eaves are indicated, be sure to include the required blocking and soffit covering (Figure 5–7). Fascia members, the pieces used to cover the exposed ends of the rafters, are taken off in lineal feet and may be 1″ or 2″ nominal thickness. If a hip roof is shown on the drawings, be certain to include solid sheathing, box cornice, and fascia on all sides of the building. Identical geometric shapes occur in roof sheathing, but be sure to include all the roof area in your take-off. Many estimators have been embarrassed by forgetting to multiply the area of one side of a simple gable roof by 2 to obtain the proper total area.

Two carpenters working together seem to make the best team or "crew" for roof framing and should be able to install about 800 BM of rafters per day. Solid plywood sheathing, ½″ or ⅝″ thick, should be installed at about 1400 sq ft per day, but if boards are used, only about one-half that production can be expected. Framing for dormers, hips, valleys, and similar locations will reduce daily output by about 10% and strip sheathing will increase the output to about 1000 to 1200 BM daily. Approximately 7 lb of 16d common nails is required for each 1000 BM of rafters and about 40 lb of 8d nails is required for 1000 BM of sheathing.

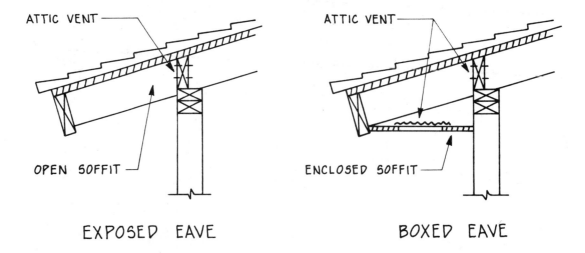

**Figure 5-7**  Eave diagrams.

**5.10 ROOF TRUSSES** With the increased number of tract houses of similar design has come the introduction of prefabricated roof trusses to replace the older job-fabricated rafter system. Trusses are essentially stronger, are easier to handle on the job, allow for other work while trusses are being built, and can be supplied in dozens of shapes and sizes. They are usually constructed of $2'' \times 4''$, $2'' \times 6''$, or $1''$ thick lumber with gang-nailed joints, or may be glued and nailed together. Whereas most wood trusses for residential use can be handled by a carpenter crew, it may be necessary to consider crane rental for large trusses or high buildings. Prices for trusses are given by the manufacturer and depend on the load to be held, pitch of the roof, span, and other design requirements.

**5.11 FINISHING CARPENTRY** Most of the foregoing information in this chapter has dealt with so-called "rough carpentry," the work that uses lumber as it comes from the sawmill and lumberyard without further machine work. Obviously, this is not the only wood material used in construction, so we must consider two other major categories; *finishing carpentry* and *millwork or cabinets*. These two areas of work use wood materials that have been shaped or otherwise finished from standard or "mill" lumber. Included in finishing carpentry are panel work, exterior siding, stairs and railings, moldings, installation of doors and windows, and similar work. Cabinetwork is simply that—fabrication of base cases, upper cases, shelving, countertops, and other types of cases. Material used for both finishing carpentry and cabinetwork may be softwoods, such as pine, fir, and redwood, or the hardwoods, such as ash, birch, maple, oak, and walnut. Considerable quantities of various plywoods, particle board, and laminated plastics are used in casework.

Also included under finishing carpentry is the *installation* of doors and windows, together with their frames and hardware. Specification and purchase of doors, windows, and finishing hardware is normally *not* a part of carpentry. In general, finishing carpentry has been the catch-all for most of the small items that are necessary to complete a project, probably because the general contractor has usually handled carpentry as his area of operation and a great many minor items have been "buy-outs," not subcontracts.

**5.12 EXTERIOR COVERING** Exterior siding or panel covering may be any one of dozens of types. "Siding" is usually solid wood in a number of patterns, $1''$ nominal thickness and $6''$ to $10''$ wide. Plywood exterior wall covering is supplied in $4' \times 8'$ sheets, $\frac{3}{8}''$ to $\frac{3}{4}''$ thick, cedar, pine, fir, or redwood faced in a number of patterns, with or without factory stain finish. Many patterns of siding or plywood are "rough sawn" to provide a more natural surface. Some exterior panels are also available in pressed hardboard or particle board with prime paint coat or plastic finish. Most exterior wood materials are installed over a layer of waterproof paper, usually 15-lb asphalt-saturated *felt*. In colder climates exterior covering may be installed over solid insulating board sheathing. Corners of siding may be mitered or covered with a cornerboard on each face.

Labor for finish carpentry must be more skilled, so a different type of workman is required and often a bit more time is required. Siding can be installed by a single carpenter, as the units are not difficult to handle, and about 250 sq ft per day is average production. Plywood or similar sheet material requires two carpenters installing 650 to 700 sq ft per day. Consider about 10% waste for all types of siding. If battens are installed over joints in a board-and-batten pattern, estimate them in lineal feet to the length required.

Waterproof building paper is used in a great number of places in any type of construction and is available from 8 lb per 100 sq ft (rosin paper) through the asphalt-saturated papers in weights of 15, 30, and 45 lb. Waterproof paper is used beneath siding, under shingles, as a base material for built-up roofing, for basement waterproofing, and in dozens of other spots. Paper is available in rolls 36″ wide and long enough to provide 108 sq ft per roll, and can be installed by one carpenter at a rate of about 3500 sq ft per day. Rigid insulating wallboard 1″ to 2″ thick is installed by one carpenter at about 750 sq ft per day. Where siding and nails are not to be painted over, all nails should be aluminum to avoid rust marks created by rain on steel nails. Siding requires about 10 to 12 lb of 6d coated nails for 1000 sq ft of siding.

## 5.13 INTERIOR CARPENTRY

Interior carpentry and finishing carpentry may be considered and estimated synonymously. Interior carpentry includes the fitting and placement of such items as paneling, moldings of all sorts, miscellaneous shelving, railings and stairs, trim at doors and windows, fireplaces, and similar items. Most of the actual material for these parts is preshaped from standard lumber in a "planing mill" or other woodworking establishment. Cabinets, counters, and built-in cases are also usually prefabricated at a mill but installed by finishing carpenters at the job site. Another large item of interior carpentry is the placement and finishing of doors, installation of pulls, locks, and other hardware on doors, windows, and cabinet work, and the proper installation of toilet and bath accessories, such as towel bars, soap dishes, and paper holders. Wood flooring is generally considered as "flooring" and is estimated and installed as a part of finishing work together with resilient flooring, carpets, and other floor coverings.

It would be virtually impossible to completely itemize labor costs for every type of interior finishing carpentry simply because the sizes and shapes may be so different. Moldings, chair rails, bases, door and window trim, handrails, and similar parts are normally installed by one carpenter at an average rate of about 250 lin ft per day. Panel work requires a crew of two carpenters, as the material is in larger sheets and installation is about 450 to 500 sq ft per day for the crew. Shelving and closet poles require considerable cutting and fitting for one carpenter and varies from 75 to 150 lin ft per day. Stair installation is a speciality and only a limited number of finish carpenters are stair builders. Stair treads are usually of $1\frac{1}{8}''$ thick oak or other hardwood and only about 15 to 18 treads and risers can be installed per day. Railings with open-spaced balusters and continuous handrails will require one carpenter's labor for 4 hr to install a normal one-story stair rail.

## 5.14 CABINET WORK

Every residence, and a large number of other buildings, requires some sort of cabinet work. The kitchen usually requires the greatest proportion of the cases, tops, and hanging shelving and may be either job fabricated or shop manufactured. Shop-fabricated cabinet work is done by skilled cabinetmakers using modern machines to cut and shape the parts and assemble them as necessary. These cabinets may be supplied completely assembled, in sets of nonassembled sections, unfinished or completely finished with all hardware in place. Job-fabricated cabinet work depends on the skill of the carpenter, who may be a "cabinetmaker" or may not. Cabinets made on the project site are often poorly made because of lack of proper machinery or lack of adequate materials of the correct kind, and are rarely "prefinished" before installation.

Lumber for cabinetwork is usually kiln dried and for cases to be paint finished are of Douglas fir, pine, or similar softwood that is readily available. For cabinets to be finished "natural," that is, with a stain and transparent coating, more exotic hardwoods may be specified. Cabinet tops are manufactured in a great variety of styles and normally use plywood as a base material. Surface finishes may be a laminated plastic, "butcher-block" wood, ceramic tile, stainless steel, or similar material. Most of these are difficult to fabricate onto the plywood base unless proper jigs and clamps are available, and this is not an easy thing to do on the project site.

Kitchen cabinets may be classed in several different groupings for cost purposes. "Base cabinets," usually about 24″ deep and 35″ high, may be drawer cabinets or supplied with shelves and doors. Wall cabinets are approximately 12″ deep by 36″ wide but may be 15″, 24″, or 30″ high, with shelves and doors. All cases should have plywood backs, although in some cheaper tract construction the cases are furnished without backs and the wall surface is exposed in the back of the case. Corner cabinets are sometimes required and are usually the same depth as adjacent cases. Base cabinets with 18″ wide drawers on top and with one door below cost about $95 to $100 in place; with four drawers but no doors, the cost is about $150 per unit; with two doors and no drawers, about $125 each. Wall cabinets 12″ deep, 15″ high, and 30″ wide with two doors cost about $50 each. Corner cabinets cost from $50 to about $75 each. Kitchen cabinets will cost an average of $75 each, exclusive of tops.

Countertops made of laminated plastic may have a coved back

splash or none, may have a square self-edge of plastic, a rolled edge of plastic, or a metal or wooden edge. Laminated plastic tops will cost about $10 to $15 per lineal foot and about $5 per lineal foot for installation. Solid laminated "butcher-block" tops of maple will cost an average of $25 per lineal foot but perhaps as much as $25 per foot for installation. Stainless steel is not a carpenter's work but when used for countertops may be estimated at $60 per square foot in place. Bathroom vanity cases will average about $150 per unit in place.

## 5.15 MISCELLANEOUS MILLWORK

Rough lumber that has been "machined" in any manner may be termed millwork and includes built-up columns, fireplace mantles, cupola work, wood louvers, trim of all sorts, and a host of other parts. Most millwork is installed unfinished, so the estimator should be certain that the same quantities that he has taken off for the millwork are carried over into the painting estimate.

Finish carpentry also includes installation of the finishing hardware to hang the doors and finish the cabinet work. Installation costs for the finishing hardware is more difficult to estimate, so a figure of ½ to 3% of the cost of the hardware may be used. Typical installation costs for lock sets will range between $12 and $15 per door or about ½ hr of a carpenter's time per door. If there are mortice locks required, this time should be increased to at least 1 hr per door. The same is true for entrance doors equipped with locks, 1½-pair butts, pulls, and night locks or dead bolts. Cabinet hardware includes hinges, catches, pulls, knobs, rim latches, and cabinet locks and should be estimated at about 100 units per day or about 8 to 10 minutes for each unit.

Bath and toilet accessories are items such as towel bars, soap dishes, paper holders, robe hooks, and similar parts. Bath and toilet accessories are *not* the plumbing fixtures, such as water closets, lavatories, tubs, and showers. These are plumbing items and are supplied and installed as plumbing work. Accessory items, like finishing hardware, are usually included in an "allowance" which is included in the contractor's bid figure but allows the owner some freedom in selection of patterns and styles that the owner may want. Typical installation costs for bath accessories is about ¼ hr of carpenter labor for each unit. Steel medicine cabinets installed in the rough opening require about 1 hr of labor time.

## PROBLEMS

**Sample problem:** Estimate the amount of lumber and plywood to complete the following structure. Floor size, 8′ × 12′, concrete floor slab. Walls 2″ × 4″ at 16″ o.c. Nine-foot plate height on high side, 8′ height on low side. Rafters 2″ × 6″ at 16″ o.c. Overhang front and rear 1′-6″, flush at other two sides. One door 3′ × 6′-8″ high, and one window 3′ × 4′-6″ high. Plywood roof sheathing, ½″ C-D grade.

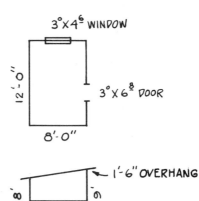

Rear wall studs 3 ea. 48″        9 pc × 8′-0″
Front wall studs 3 ea. 48″        9 pc × 10′-0″
Side wall studs 3 ea. 48″        6 pc × 10′-0″ × 2  12 pc
    deduct for door and window  4 pc × 10′-0″
Corners 4 corners 2 pc  ea.    8 pc  × 10′-0″
Cripples etc. @door/window  8 pc  × 10′-0″
Plates; bottom 12 × 2  8 × 2    40 lin ft    2 pc × 8′-0″
    top frnt 12 × 2 × 2        48 lin ft    4 pc × 10′-0″
    top side 10 × 2 × 2        40 lin ft    4 pc × 12′-0″
Plate blkg        12 × 2        24 lin ft    2 pc 2 × 6 × 12-0″
Rafters 3 ea 48″ plus end                10 pc 2 × 6 × 12′-0″
Bracing  4 pc  1 × 6 × 10′-0″
Lintel @door/window 1 pc  4 × 6 × 8′-0″

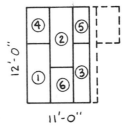

PLYWOOD REQ'D

### Lumber list

| 4 pc  | 1 × 6—10-0  | 20.0 Board Measure |
|-------|-------------|--------------------|
| 11 pc | 2 × 4— 8-0  | 58.6 |
| 30 pc | 2 × 4—10-0  | 199.8 |
| 6 pc  | 2 × 4—12-0  | 48.0 |
| 2 pc  | 2 × 6—12-0  | 24.0 |
| 1 pc  | 4 × 6— 8-0  | 16.0 |

364.4 Board Measure

5 pc  plywood 48 × 96 × ½

**5-1.** Make a take-off of lumber required to rough frame the sub-floor of a free-standing building 20′ × 24′ (see Figure 5-2). Use 4″ × 6″ girders at 5′ the long way of the building, 2″ × 6″ floor joists at 16″ o.c., and ¾″ plywood subfloor.

**5-2.** Estimate the material required to install the subfloor indicated in the house plans included in this book (beginning on page 194). Make a proper lumber list of these requirements.

**5-3.** Make a take-off of all rough framing required to completely frame the same house. Provide a proper lumber list.

# chapter

# 6

# Roofing
# and
# Waterproofing

Roofing, as a generic term, includes many different kinds of coverings that may be installed to keep rain, snow, or other weather conditions from entering a building. Most residential roofing is either shingles or shakes manufactured from wood; composition shingles manufactured from asphalt, fiberglass, or combinations; or in lesser quantity, various tile shapes manufactured from clay or concrete. In a relatively few instances roofs are of metal or are built up with alternating layers of tar or asphalt and felts, or in some locations of slate split into shingles. Regardless of the material used for roofing there is also a quantity of sheet metal (copper, galvanized steel, zinc) used as flashing to helps weathertight a roof installation, together with some plastic roofing compounds or sealers, and a variety of roof accessories.

Roof materials may be installed over solidly sheathed surfaces or upon "skip-spaced" nailers. Solid roof sheathing of plywood may be installed rapidly, forms a good diaphragm, and provides a good working surface. Such solid surface is absolutely necessary for composition roofing to maintain its shape without forming small ridges and valleys. Spaced roof sheathing or "skip spacing," using $1'' \times 4''$ material at $8''$ o.c., is often used when wood shingles or shakes are used, as these materials are rigid enough to span the openings between boards.

Shingles, shakes, and composition shingles should not be installed on roofs with a pitch of less than 3 in 12. Roll roofing may be installed on roofs from dead flat to 4 in 12, but extreme care is needed that roll roofing is adequately mopped in place or seams sealed to provide a watertight barrier.

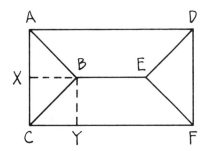 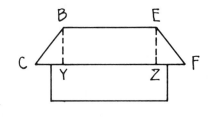

**Figure 6-1** Estimating roof areas.

## 6.1 ESTIMATING ROOF AREAS

Roofing, regardless of type, is estimated by the *square,* representing 100 sq ft of roof area. This area may be 10′ × 10′, 2′ × 50′, or any other combination. The cost of materials as well as cost of labor is determined primarily by the shape or pitch of the roof, openings in the roof plane, and distance above the ground. To find the area of a flat roof is relatively simple: multiply width by length. Pitched roofs require that a reasonable determination of the slanted length be calculated from the elevation drawings (see Figure 5–6).

The area of plain gable roof or a shed roof is a matter of multiplying the roof pitch (rafter) length by the ridge or eave length. In the shed roof this will give the area required, but with a gable roof this area must be multiplied by 2 to obtain the total roof surface. To find the area of a hip roof, consider that the surface is composed of a number of triangles and oblongs. Refer to Figure 6–1. The hip triangles may be calculated as triangle ABC, then multiplied by 2 to obtain area for both ends. Side areas require rafter length BY multiplied by ridge length BE to obtain an oblong area BEYZ, *plus* BY multiplied by CY, for a total of one side. This area multiplied by 2 *plus* the area of the two end triangles will give the total area. An alternative method is to add BE to CF and multiply by BY for the area of *both* sides, then add this answer to the end triangles for total area. In any case, be sure to include *both* sides and *both* ends. Divide the total area by 100 to find the number of squares required.

Areas of other roof planes may be determined by calculating various geometric shapes, such as triangles, arc of circles, and the more mundane squares, oblongs, or circles. Do not be reluctant to sketch the various planes. Such sketches quite often will point out an easy solution for complicated roofs. Do not fail to add the various areas together for total area. It is not uncommon for estimators to fail to do this, with a resulting disastrously low bid, so *double check.*

## 6.2 SHAKES AND SHINGLES

Shakes and shingles are manufactured from wood, normally from cedar or redwood. In the early manufacture of these products they were hand-split but now are usually produced by machines. The major association of producers of wood shingles and shakes is the Red Cedar Shingle and Hand-Split Shake Bureau, which establishes the size, grade, and so on. Lengths of shingles vary from 16″ to 24″, available in four grades from clear to undercoursing, and in butt thickness from 0.40″ to 0.50″ Shingles are provided random width from 2″ to 12″ and about four bundles are required per square of roofing. When shingles are used for wall covering, approximately two bundles per square are required. Normal exposure for roof shingles is from 3½″ to 5½″ and about 10% additional should be allowed for double coursing at eaves, cutting at ridges and valleys, and general waste. Labor on plain gable or hip roofs varies from 10 bundles to 15 bundles per day, depending on whether carpentry labor or shingler labor is employed. About 4 lb of 3d or 4d nails is needed per square. For quick estimating, wood shingles, installed 5″ to weather, cost about $150 per square in place, with the price of #1 shingles at about $75 per square.

Wood shakes are longer and heavier than shingles and give a more massive and rustic appearance, which may be desirable in some designs. Shakes are usually supplied with sawn backs but with rough split faces and vary in length from 18″ to 24″ with butt thickness from ½″ to 1¼″ and weather exposure from 5½″ to 10″. Coverage of one square of roof (100 sq ft) installed at 10″ exposure requires five bundles, or seven bundles when exposed at 7″ to weather. Installed costs vary in the range $175 to $200 per square, with the cost of shakes about $100 per square.

Two methods of installation of either shingles or shakes are common. When skip sheathing is used, the shingles are simply nailed to the sheathing strips, spanning the intermediate space, and in most cases without waterproof paper (felt) between layers. When installed over solid sheathing the upper end of each course or layer is covered with a strip of waterproof paper. In either case each shingle is fastened with only two nails located, so they are concealed by the next course and each shingle is spaced ¼″ to ⅜″ from the next unit. Shingles are laid dry and when wet by rain expand across the grain to fill these spaces and provide a solid roof surface. The first course at the eaves is usually doubled or tripled and all joints between successive courses are staggered. A good shingle or shake roof will give excellent service for at least 20 years, and many installations are more than 50 years old and still weathertight.

## 6.3 COMPOSITION ROOFING

Most composition roofing is manufactured from asphalt with other materials, such as fiberglass, stone aggregate, or plastics. Material is available in rolls or as cut strip shingles, as indicated in Figure 6–2, and is installed on solid roof sheathing with nails, hot asphalt, or roofing cement. Roll roofing, which may be asphalt-saturated paper or coal-tar saturated, is manufactured in 15 to 65 lb per square, black in color, and without further surface preparation. Such material is *not* suitable for exposed roof surfaces but is extensively used as a moisture barrier in connection with other types of roofing materials, as a backing for plaster work, as a sheet material for waterproofing, and as a vapor barrier beneath wood siding and in other locations.

| TYPE | PRODUCT | SHIP WT. SQ. (LB) | PKG. PER SQ. | LENGTH | WIDTH | UNITS PER SQ. | SIDE LAP | TOP LAP | HEAD LAP | EXPO-SURE |
|---|---|---|---|---|---|---|---|---|---|---|
| ROLL ROOFING | SATURATED FELT | 15 | 1/4 | 144' | 36" | | 4" TO 6" | 2" | | 34" |
| | | 30 | 1/2 | 72' | 36" | | 4" TO 6" | 2" | | 34" |
| | SMOOTH | 65 | 1 | 36' | 36" | | 6" | 2" | | 34" |
| | | 55 | 1 | 36' | 36" | | 6" | 2" | | 34" |
| | | 45 | 1 | 36' | 36" | | 6" | 2" | | 34" |
| | MINERAL SURFACE | 90 | 1 | 36' | 36" | 1.0 | 6" | 2" | | 34" |
| | | 90 | | | | 1.075 | 6" | 3" | | 33" |
| | | 90 | | | | 1.15 | 6" | 4" | | 32" |
| | PATTERN EDGE | 105 | 1 | 42' | 36" | | | 2" | | 16" |
| | | 105 | 1 | 48' | 32" | | | 2" | | 14" |
| | 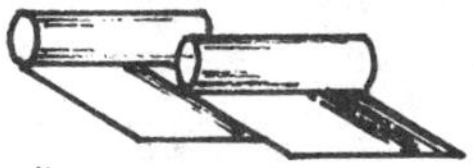 19" SELVAGE EDGE | 110 TO 120 | 2 | 36' | 36" | | | 19" | 2" | 17" |
| STRIP SHINGLE | 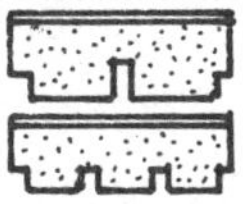 SQUARE BUTT | 235 | 3 | 36" | 12" | 80 | | 7" | 2" | 5" |
| | 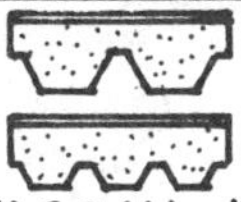 HEXAGONAL BUTT | 195 | 2 | 36" | 11 1/3" | 86 | | 2" | 2" | 5" |
| INDIVIDUAL |  LOCK | 145 | 2 | 16" | 16" | 80 | 2 1/2" | | | |
| |  STAPLE | 145 | 2 | 16" | 16" | 80 | 2 1/2" | | | |
| GIANT INDIVIDUAL | 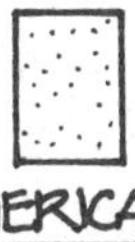 AMERICAN | 330 | 4 | 16" | 12" | 226 | | 11" | 6" | 5" |
| | 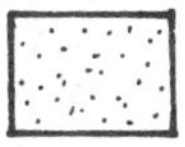 DUTCH LAP | 165 | 2 | 16" | 12" | 113 | 3" | 2" | | 10" |

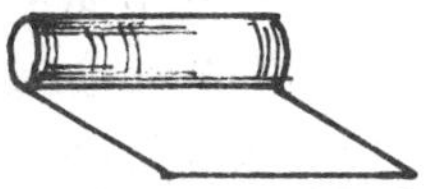

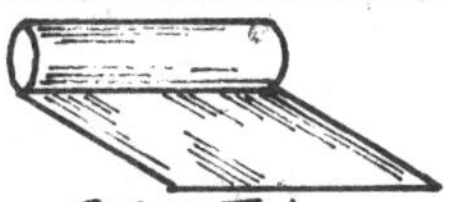

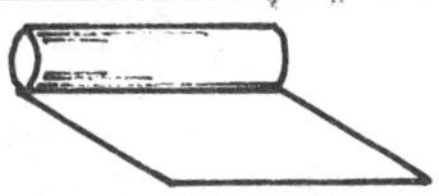

**Figure 6-2** Asphalt roofing products.

The most common type of roll roofing is "mineral surfaced": heavy-weight saturated paper (felt) which has finely ground colored mineral particles rolled into the top surface which is exposed to the elements. This material may be installed with or without additional underlayment of other felts and is usually spot-mopped in place and end and edge laps sealed with hot asphalt. This type of installation is not recommended for roofs flatter than 1 in 12 and cannot be expected to have a useful life span beyond 5 years. The cost of material is approximately $8 per square and labor for an installation crew about $4 per square, for a total of about $12 per square in place. Life expectancy may be improved by the addition of layers of smooth-saturated underlayment mopped in place at a cost of about $8 per square per layer. In cases of dead-flat installation a built-up roof of several layers may be a better investment (see Section 6.4).

Composition shingles are of two major types: strip shingles approximately 36″ long and including several shingle units, or individual shingles of various designs. Strip shingles are heavy weight, have a variety of surface textures on the exposed portions, and the mineral coating is usually colored. They are installed with overlapping edges which have a heat-sealing adhesive, and are nailed at the upper edge, which is concealed by the next course. Individual shingles are available in a wide range of weights, shapes, and surface textures, and are usually nailed in place. Both strip shingles and individuals require a solid roof deck and are most often applied over one or more layers of waterproof felt underlayment. Special care should be exercised to ensure that the lower edges of composition shingles are securely fastened with self-sealing adhesive, roofing cement, hot asphalt, or nailing, as wind will curl up or tear off shingles if they are not adequately retained flat. Composition shingles should give at least 10 years of service and cost about $85 to 90 per square for 240-lb class A fire-rated shingles in place.

## 6.4 BUILT-UP ROOFING

Flat roofs on buildings are always a problem and a great number of contractor "call-backs" could be avoided if only a little pitch for drainage was installed in flat roofs. Several factors contribute to the flat-roof drainage problem. In many cases roofs are designed flat to provide a level ceiling line across the bottom of the rafters. With any reasonable span there will be some slight deflection of the rafters downward, and this occurs in midspan. If rafters are slightly undersized this condition is worsened, so that any additional load of roof sheathing and roofing material only increases the deflection. Flat roofs will eventually drain over the edge when enough rainwater collects but before this happens the weight of the additional rain collecting at midspan creates more deflection, which allows more water to pond at midspan, which causes more deflection, and so on. As little as ¼″ per foot pitch could eliminate this condition and can easily be done by tapering the rafters, adding tapered strips to the upper side of rafters, by pitching roof insulation, or designing the flat roof slightly out of level.

**Table 6-1**  Specifications requirements for 20-year bonded roof

|  | Weight (lb) |
|---|---|
| 1 layer 30-lb base felt | 30 |
| 3 layers 15-lb intermediate felts | 45 |
| 1 spot mopping of base sheet | 8 |
| 3 moppings between 15-lb felts | 75 |
| 1 flood coat asphalt | 50 |
| ⅜″ finish gravel at 400 lb/square | 400 |
| Total weight per square | 608 |
| Weight per square foot | 6.08 |

Flat roofs must be considered as pools in most cases and the roofing must be installed watertight above the roof level to the flashing. The "built-up" roof is the most common answer and is composed of three to five layers of saturated felts, each layer embedded in hot asphalt or hot coal tar, and the finished surface flood mopped with hot asphalt or hot coal tar and adequately covered with gravel or crushed rock. Such roofs are often referred to as "15-year bonded" (three layers of waterproof felts) or as "20-year bonded" (four or five layers of waterproof felts). When properly installed in accordance with the material manufacturer's requirements, including flashings, the manufacturer will then "bond" the installation for a small premium to ensure its weathertightness for the time period agreed on. Installation of built-up roofs requires special skills and special equipment so is not work for amateurs but is also not often used in residential construction. The specifications requirements for a 20-year bonded roof, given as Table 6–1, should indicate that such a roof also is not light weight. The cost for such a roof, using four journeymen, two helpers, and a foreman, will run about $90 per square in place, plus $165 for material. The crew should be able to apply about 18 to 20 squares per day over wood or concrete decks.

## 6.5 OTHER ROOFS

In addition to the types of roofing already discussed, there are a few that may not be as familiar. Some are continuations of roofing used decades ago, whereas others are quite recent. Prominent among these earlier materials are clay tile roofs and slate shingles, while elastomeric plastic roofs are modern, and metal roofs appear somewhere in the middle. Use of these materials is somewhat restricted by building design, geographic location, weather conditions, and the availability of materials as well as the desires of the owner or designer.

Clay tiles were originally of only one shape, the so-called "Spanish" or "Mission" type, which were formed over the upper leg of the maker. This general shape has been retained in the modern tile, which is tapered, about 13″ to 14″ long, half circle in section, ½″ thick, and evenly fired for a uniform color or "flash-fired" for a variation of the normal red color. From this start there have been a number of variations, including interlocking tiles in flat shape or multiples of mission tile. Tiles are now manufactured of clay, cement, asbestos cement, plastic compounds, and various combinations and in a great variety of colors. Originally, clay tiles were installed over poles or strip sheathing, then over waterproof paper on solid sheathing, and today most tiles are installed over solid decks. Clay tiles cost about $185 per square in place, while concrete tiles average about one-half that amount. Installation time is about 4½ hr per square for clay tiles and about 3 hr for concrete tile. If waterproof paper or other material is to be used beneath tile, and this is usually the situation, be sure to include the cost of that installation in addition to the cost of the tile installation.

There is still considerable use of slate shingles for roofing of restorations and "classic"-designed buildings in the United States, particularly in the Eastern area. Slate is a natural stone, black to blue-gray in color, supplied in nominal sizes from 6″ × 10″ to 14″ × 26″ and in ³⁄₁₆″ to ¾″ thickness. Shingles are punched for two nails each and all slate is installed over 30-lb waterproof paper, with 3″ minimum head lap and staggered joints in each course. Copper or brass fastenings are usually require. Most slate roofs occur in classically-designed buildings, with dormers, chimneys, and other interruptions in the roof plane so at least 10% should be added to the estimated roof area to provide adquate material. A roofing crew of two installers and one helper should lay approximately 400 sq ft (4 squares) in an 8-hr day at a cost of $300 per square, including the 30-lb underlayment. Properly installed and maintained slate roofs should last the life of the building.

In recent years there have been a considerable number of sheet or sprayed roofing materials devised from various plastic compounds. For the most part they are used primarily in commercial or industrial construction, but occasionally they may be advantageously used in larger residences, apartments, or condominiums. These materials are basic polyurethane elastomeric compounds available in sheet form or liquid, cold applied, from 6 to 18 mils thick, black or colored, and are usually applied by manufacturer-authorized representatives. Liquid application has the advantage that, properly applied, there are no seams or laps to consider and most products may be applied on vertical as well as horizontal surfaces. This type of roofing costs about $180 per square installed with about $90 for materials and requires about 3½ hr of labor per square.

## 6.6 SHEET METAL WORK

Sheet metal work occurs in a considerable number of forms in construction: as duct work in air conditioning, as metal siding, as fabricated tables, benches, or other fixtures, and as flashings of all types. The metal used may be aluminum, copper, galvanized steel (GI—galvanized iron), lead, stainless steel, or zinc. Comparable thicknesses of various sheet metals are given in Table 6-2. Any and

**Table 6-2**  Comparable sheet metal thickness

| Ga. | Galv. steel | | Copper | | | Aluminum | | Stainless steel | | Zinc | | Lead | |
|---|---|---|---|---|---|---|---|---|---|---|---|---|---|
| | Thickness (in.) | Lb/sq ft | Oz. | Thickness (in.) | Lb/sq ft | Thickness (in.) | Lb/sq ft | Thickness (in.) | Lb/sq ft | Thickness (in.) | Lb/sq ft | Thickness (in.) | Lb/sq ft |
| 10 | 0.1345 | 5.78 | 96 | 0.1250 | 6.00 | 0.1250 | 1.76 | 0.1250 | 5.25 | — | — | 0.1250 | 8.00 |
| 11 | 0.1196 | 5.16 | 88 | 0.1190 | 5.50 | — | — | — | — | — | — | — | — |
| 12 | 0.1046 | 4.53 | 72 | 0.0912 | 4.50 | 0.1094 | 1.44 | 0.1094 | 4.59 | — | — | 0.0937 | 6.00 |
| 13 | 0.0897 | 3.91 | 64 | 0.0897 | 4.00 | 0.0907 | 1.28 | — | — | — | — | — | — |
| 14 | 0.0747 | 3.28 | 56 | 0.0755 | 3.50 | 0.0781 | 1.10 | 0.0781 | 3.28 | 0.0700 | 2.62 | — | — |
| 15 | 0.0637 | 2.97 | 48 | 0.0647 | 3.00 | 0.0641 | 0.91 | — | — | — | — | 0.0625 | 4.00 |
| 16 | 0.0598 | 2.66 | 44 | 0.0593 | 2.75 | — | — | — | — | — | — | — | — |
| 17 | 0.0538 | 2.40 | 40 | 0.0539 | 2.50 | — | — | 0.0563 | 2.37 | 0.0500 | 1.87 | — | — |
| 18 | 0.0478 | 2.16 | 36 | 0.0485 | 2.25 | 0.0469 | 0.66 | — | — | — | — | 0.0468 | 3.00 |
| 19 | 0.0418 | 1.90 | — | — | — | — | — | 0.0438 | 1.84 | — | — | — | — |
| 20 | 0.0359 | 1.65 | 28 | 0.0377 | 1.75 | 0.0359 | 0.51 | 0.0375 | 1.58 | 0.0360 | 1.35 | — | — |
| 21 | 0.0329 | 1.53 | 24 | 0.0323 | 1.50 | 0.0320 | 0.45 | — | — | 0.0320 | 1.20 | — | — |
| 22 | 0.0299 | 1.41 | — | — | — | 0.0313 | 0.44 | 0.0313 | 1.31 | — | — | 0.0312 | 2.00 |
| 23 | 0.0269 | 1.28 | 20 | 0.0270 | 1.25 | — | — | — | — | — | — | — | — |
| 24 | 0.0239 | 1.16 | 18 | 0.0243 | 1.13 | — | — | 0.0250 | 1.05 | 0.0240 | 0.90 | 0.0234 | 1.50 |
| 25 | 0.0209 | 1.03 | 15 | 0.0202 | 0.94 | 0.0201 | 0.28 | 0.0219 | 0.92 | 0.0200 | 0.75 | — | — |
| 26 | 0.0179 | 0.91 | 13 | 0.0175 | 0.82 | 0.0179 | 0.25 | 0.0172 | 0.72 | 0.0186 | 0.67 | — | — |
| 27 | 0.0164 | 0.84 | 12 | 0.0162 | 0.75 | — | — | — | — | — | — | — | — |
| 28 | 0.0149 | 0.78 | 11 | 0.0148 | 0.68 | 0.0142 | 0.20 | 0.0156 | 0.66 | 0.0140 | 0.52 | 0.0156 | 1.00 |
| 29 | 0.0135 | 0.72 | 10 | 0.0135 | 0.63 | — | — | — | — | — | — | — | — |
| 30 | 0.0120 | 0.66 | 9 | 0.0121 | 0.56 | 0.0100 | 0.14 | 0.0125 | 0.53 | 0.0120 | 0.45 | 0.0117 | 0.75 |

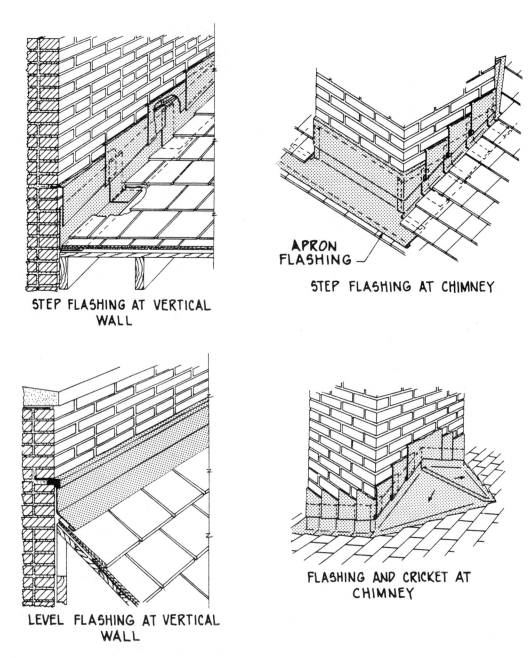

STEP FLASHING AT VERTICAL
WALL

APRON
FLASHING

STEP FLASHING AT CHIMNEY

LEVEL FLASHING AT VERTICAL
WALL

FLASHING AND CRICKET AT
CHIMNEY

**Figure 6-3** Roof flashings *(from Architectural Sheet Metal Manual,* 3rd. ed., 1979, with permission of Sheet Metal and Air Conditioning Contractors National Association–SMACNA).

all these may be used in connection with roof installations as flashings, gutters, valley material, downspouts, or similar parts. Some sheet metal is manufactured, fabricated, and paid for by the gauge of the metal, some by weight, some by inch thickness. Installation is generally controlled by the unions, not by state licensing organizations, and in some parts of the United States sheet metal parts may be fabricated and installed by roofers, whereas in other areas sheet metal work is a separate trade and fabrication and/or installation is not done by roofers.

**Table 6-3** Costs for flashings and gutters

| | Material | Total | Output (1 man/8 hr) |
|---|---|---|---|
| *Flashings (cost/sq ft)* | | | |
| Aluminum, 0.040 " | $0.70 | $1.90 | 140 |
| Copper, 16 oz | 1.10 | 3.20 | 100 |
| Galv. steel, 24 gauge | 0.95 | 2.10 | 150 |
| Stainless steel, 0.025 " | 2.25 | 3.60 | 150 |
| Zinc alloy, 0.040 " | 1.60 | 3.00 | 150 |
| *Gutters (lin ft)* | | | |
| Aluminum, 5 " × 0.027 " | 0.60 | 2.10 | 120 |
| Copper, 5 " × 16 oz | 2.00 | 3.75 | 120 |
| Galv. steel, 5 " × 26 gauge | 0.70 | 2.25 | 120 |
| Stainless steel, 5 " × 0.025 " | 3.00 | 5.00 | 120 |

Several roofing items are fairly common in residential construction and may also occur in other types of building. Probably two of the most repetitive are flashings between horizontal surfaces and vertical surfaces, and metal valleys where one pitched roof surface joins another. The first is accomplished by using formed metal strips where roofs are reasonably flat or by using a series of bent plates installed step-fashion at pitched locations, as indicated in Figure 6–3. The upper edges of these flashings is terminated in a reglet in concrete or masonry or may be sealed in a mortar joint or waterproofed with roofing compound. Valley material is supplied in lengths up to 10 ft, often has an inverted vee shape at midpoint of the valley to channel the runoff, and has flanges adequate to extend beneath the adjacent roofing at least 6 ". Most flashing and gutter material is from 22 to 26 gauge in galvanized steel or equivalent in other metals. Cost of material is the greatest factor in determining the cost of roofing sheet metal items, as indicated by Table 6-3. Downspouts for gutters depend on the area to be drained for size, but 3 " × 4 " will cost about $2.50 per lin ft in aluminum, $4.00 per lin ft in copper, and about $2.25 per lin ft in 26-gauge galvanized steel.

## 6.7 METAL ROOFING

Metal roofs for residential construction may also be classified as a "special" roof and are not common in occurrence. Metal roofs may be one of two major forms: roll or sheet stock fabricated on the roof surface, or metal shingles applied in a manner similar to wood shingles. Roll stock may be aluminum, copper, galvanized steel, lead, or other alloys, while shingles are generally of aluminum or galvanized steel with a weather-resistant colored surface coating.

Metal roofs may be flat seam, standing seam, or batten seam construction (Figure 6–4). With all metal roofs there must be an underlayment of waterproof paper, usually 30 lb per square. Recommended thickness of metal is 0.032 " for aluminum, 26 or 24 gauge for galvanized steel, and 16 to 20 oz for copper. The flat-seam method provides pans sealed or soldered together without projec-

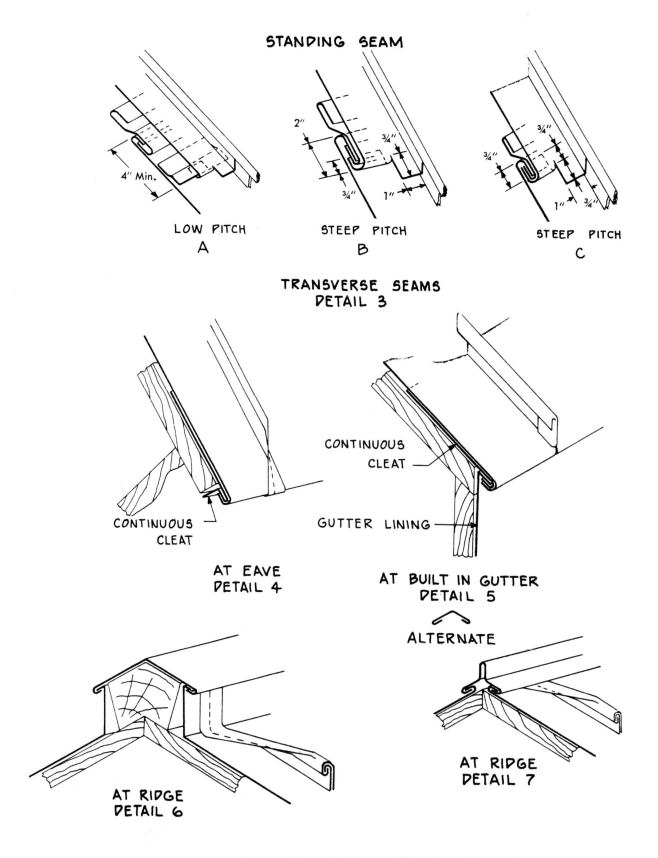

**Figure 6-4** Roof seams (*from Architectural Sheet Metal Manual,* 3rd. ed., 1979, with permission of Sheet Metal and Air Conditioning Contractors National Association–SMACNA).

BATTEN SEAM

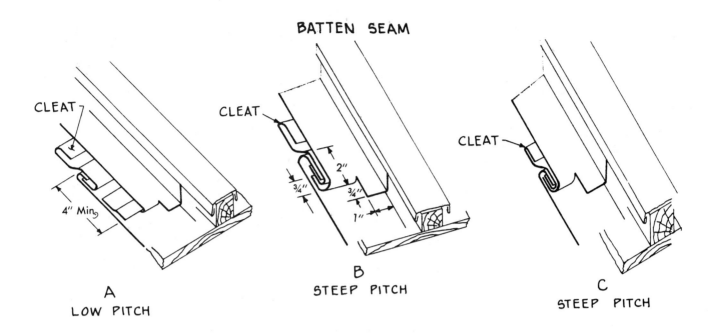

CLEAT

4" Min

A
LOW PITCH

CLEAT

2"

¾"

¾"

1"

B
STEEP PITCH

CLEAT

C
STEEP PITCH

TRANSVERSE SEAM
DETAIL 4

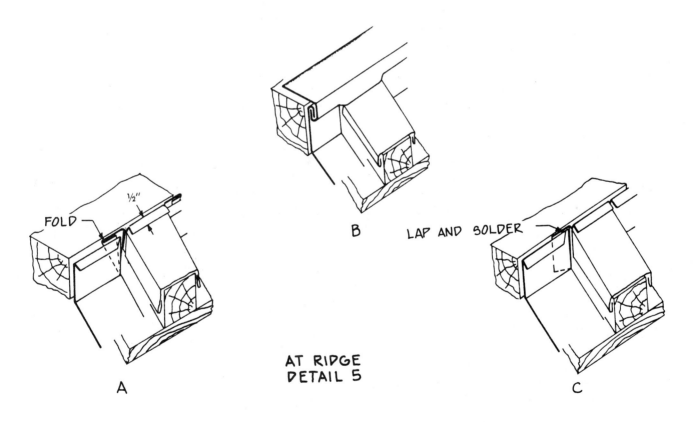

½"

FOLD

A

B

LAP AND SOLDER

C

AT RIDGE
DETAIL 5

**Figure 6-4** (Continued.)

tions greater than twice the thickness of the metal used and may be used on very low roof slopes. Standing seam is used on roofs with a pitch of 3 in 12 or greater and uses seams formed by overlapping the metal of adjoining sheets or pans. Batten seam roofs have the seams formed over wooden battens spaced at not more than 20″ on center. Careful forming of pans and joints allows for expansion and contraction and provides a surface with a minimum of exposed fastenings. Cost of batten seam work is most expensive, in the range of $500 per square for copper/zinc alloy, $350 per square for 16-oz copper, and about $200 for galvanized steel. Standing-seam work is about 2% less expensive and flat-seam work is about 3% less expensive.

Enameled or porcelain-coated metal shingles are now available in many shapes and colors as individual units or multiple strips. Most, however, are pressed from 0.020″ aluminum with a rustic wood-appearing surface or a smooth surface, are available in lengths of 36″ and 48″, 10″ nominal width, and are formed to interlock at edges and ends. Most of these metal shingles will meet requirements for class B fire rating when installed over one or more layers of 30-lb felt, and are designed to resemble a thick-butt wood shingle. Weight is about 40 lb per square and hip caps, eave starter strips, ridge caps, and companion parts are available. Although nearly every color imaginable is available, most residential use is of deep-tone wood colors, with other colors used for commercial installations on mansard projections or building fronts. Shingles in place cost about $150 per square, with ridge caps, valleys, and similar parts at about $3 per lineal foot in place. Some shingles have factory-applied insulation backings, which adds $15 to $20 per square to material costs.

## 6.8 WATERPROOFING AND DAMPPROOFING

Waterproofing is kindred to roofing, and is usually installed by roofers, but relates to stopping water from entering a building through vertical surfaces or other surfaces not normally considered roofing. The most common requirement for waterproofing is application on the exterior surfaces of basement walls, especially walls made of masonry. Other uses are as "pans" in shower or toilet rooms beneath tile, or as water stops integral with basement floors when known underground water conditions exist, and for linings of plant boxes and similar receptacles.

Waterproofing may be installed in three major forms: as multiple layers of asphalt-saturated building paper combined with hot asphalt, as a troweled-on mastic, or as single or multiple layers of plastic or rubberized sheet in cold adhesive. As most waterproofing is concealed almost immediately after it is placed, it is particularly important that it be done properly. In Figure 6–5 some suggestions are illustrated that will assure a dry basement. Note that waterproofing is carried below the floor grade and is sandwiched in the floor slab. A layer of paper-pulp board is used over the waterproofing to protect it from damage during back-filling operations,

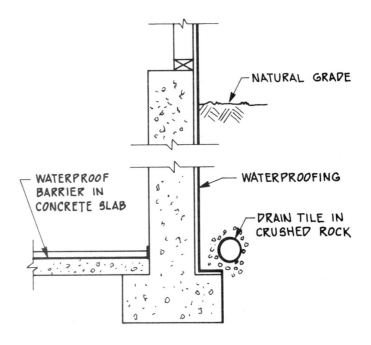

**Figure 6-5** Basement wall waterproofing.

and a drain system is shown to remove water before it reaches the building or as it is drained off the waterproofing. Cement parging, a ½″-thick layer of cement plaster, is often used as a water barrier but is not applied by roofers but by plasterers and costs about $2.50 per square foot in place.

Waterproofing is estimated in square feet or hundreds of square feet. Brushed-on or sprayed-on bituminous coatings will cost about $0.50 per square foot two coats and about $0.10 per square foot more when troweled on. When 30-lb building paper is installed in double layers with hot asphalt as an adhesive, the cost will approximate $8 per 100 sq ft, with the cost of two-ply installation of 0.002″-thick polyethylene about the same. Elastomeric waterproofing ⅟₃₂″ thick will cost about $100 per 100 sq ft, and neoprene rubber sheet, nylon reinforced, ⅟₃₂″ thick, about $150. Membrane waterproofing on slabs, two-ply felt, costs about $85 per 100 sq ft, or if fluid neoprene is applied to 50 mils thick, the cost will approximate $180 per 100 sq ft. Protective board ½″ thick, asphalt coated, installed in mastic at the rate of 450 sq ft per day, costs about $0.50 per square foot.

Waterproofing is fairly easily calculated for area, as most is taken from the drawings and is in simple geometric shapes. Some attention should be made to requirements for placing material at the bottom of footings by either adding 10% to the quantity or adding to labor only. In addition, there is about 10 to 20% natural waste and spillage when using hot asphalt or cold adhesive and a smaller amount for sheet materials. Protective boards are usually supplied in standard 4′ × 8′ sizes, so again there may be waste due to cutting of boards to correct size.

## 6.9 CAULKING AND SEALANTS

Caulking and sealants are a sort of catch-all to ensure that there are fewer air or weather leaks in a building, and the responsibility for proper installation of caulking and sealants may be spread through a number of trades. Large cracks are filled with polyethylene rod from ¼" to 1" in diameter at a cost of $35 to $40 per 100 lin ft. Gun-grade caulking material is then installed over the rod and is available in bulk or cartridge form in acrylic, butyl, latex, polysulfides, polyurethanes, or silicone rubber and costs vary from about $0.60 to $0.90 per lineal foot in place. Neoprene or polyvinyl chloride closed-cell gasket sealers cost about $1.25 per lineal foot for ½" × 2" to $5.00 for the ½" × 12" size.

Calculation of the amount of caulking required is not particularly easy or accurate since there may be many spots that need to be closed due to misfits or changes in the construction or the part used. Most caulking is at doors and windows, so a rule of thumb may be used to allow 18 to 20 lin ft of caulking for each opening, and a workman should be able to caulk about 30 openings per day, using material at about 150 lin ft per gallon.

### PROBLEMS

**Sample problem:** Estimate the quantity of roofing required for a gable roof with gable intersecting wing at right angles. The main portion is 16'-0" wide eave to eave, plate height 8'-0", ridge length 36"-0". The wing section is 12'-0" wide, plate height, 8'-0", ridge length 14'-0". Overhang at all eaves is 2'-0". The pitch of the roof is 4 in 12.

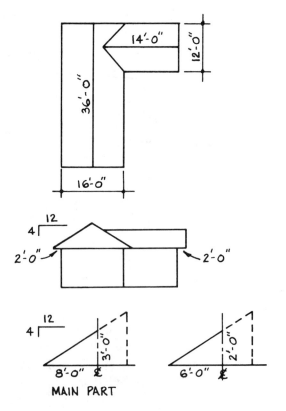

MAIN PART

Main portion:  length of slope $\qquad$ $8^2 + 3^2 = \sqrt{?}$

$\qquad\qquad\qquad\qquad\qquad$ $64\ 9\ 64 + 9 = \sqrt{73} = 8'\text{-}4''$

Area of roof:  $36' \times 8'\text{-}4(8.33) \times 2 = 549.78$ sq. ft.

Wing portion:  length of slope $\qquad$ $6^2 + 2^2 = \sqrt{?}$

$\qquad\qquad\qquad\qquad\qquad\quad$ $36 + 4 = \sqrt{40} = 6.66'$

Parts A  B $6.66 \times 3 \times 2 = 39.96$   This can be added to total but then is again subtracted as double coverage will result—disregard but reduce wing ridge to $9'\text{-}0''$   $9 \times 6.66 = 119.88$ sq  ft

Main roof  +  549.78 sq  ft

Wing roof  +  119.88 $\qquad$ $\dfrac{669.66 \text{ sq.ft}}{100 \text{ sq.ft.}} = 6.7$ sq

$\qquad\qquad$ 669.66 sq  ft

$\qquad\qquad\qquad\qquad\qquad$ Waste and 10%  $\underline{\quad .6\quad}$

$\qquad\qquad\qquad\qquad\qquad$ Requires $\qquad$ 7.3 sq

$\qquad\qquad\qquad\qquad\qquad$ SAY 7.50 squares

**6.1** Estimate the quantity of roofing required for a shed roof, high side $14'\text{-}0''$, low plate height $9'\text{-}0''$, slope $2\frac{1}{2}$ in 12, overhang front and rear $2'\text{-}0''$, eave length $50'\text{-}0''$.

**6.2** Using the house plans included in this book, pages 194-197, calculate the lineal feet of metal gutters required at all eaves.

**6.3** What is the in-place cost for a shingle roof for the included house plan using 5/2 wood shingles exposed $5\frac{1}{2}''$ to weather?

# chapter

# 7

# Insulation and Acoustics

Insulation and acoustics are often confused with each other as to their function. Insulation is concerned with temperature, repelling heat or cold, or containing it. Acoustics is related to sound, usually absorbing it, but also possibly reflecting it. Either or both may be included in any residential work, but acoustic installations are more common in commercial work. Insulation is available as loose fill, rigid board, flexible batts, reflective foil, or plastic foam. Acoustics and noise control are similar, but acoustics generally refers to surface applications, while noise control may be separation of the framing, padding of parts to minimize vibration, or arrangement of construction units.

## 7.1 INSULATION

Insulation is not simply a semisolid material used to retard transfer of heat or cold but includes consideration of vapor barrier, air spaces, reflective surfaces, and the thickness of the insulating material used. In wall construction the flexible batts of fiberglass or other mineral fiber, covered each face with paper or foil, is probably the most commonly used, with asphalt-treated rigid board used as a sheathing material a reasonable second. Loose fill in the form of pellets or small granules of mineral wool or similar material is also used extensively in construction. Folded foil in many designs and blown plastic foam are used in relatively small quantities. The value of insulation is a relationship among thickness, vapor barrier, and air space and is expressed as "R value." The typical requirements for R value for the United States are shown in Figure 7-1. All insulation now has the R value clearly indicated by the manufacturer.

Location of insulation in a wood-framed building is not very

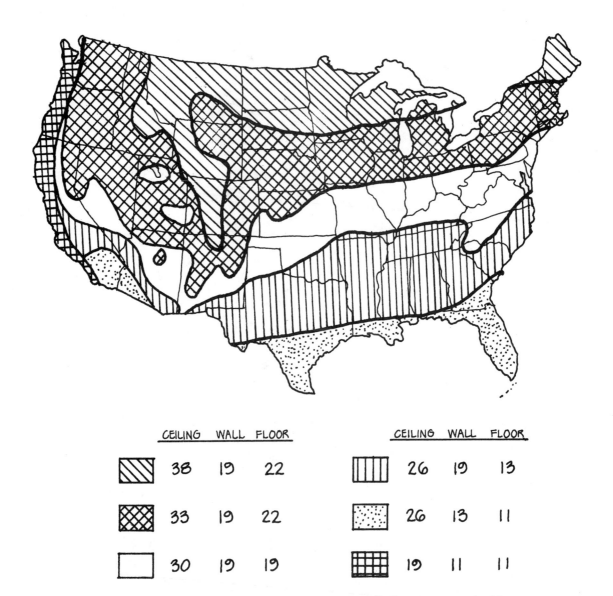

| CEILING | WALL | FLOOR | | CEILING | WALL | FLOOR |
|---------|------|-------|--|---------|------|-------|
| 38 | 19 | 22 | | 26 | 19 | 13 |
| 33 | 19 | 22 | | 26 | 13 | 11 |
| 30 | 19 | 19 | | 19 | 11 | 11 |

**Figure 7-1**   Recommended R values for the United States.

complicated. The three principal locations for insulation are in the floor, walls, and ceiling or roof structure. Batt insulation is fabricated for a pushed fit between studs or joists located at 16″ o.c. and side tabs are nailed or stapled to the wood members. Care should be exercised to ensure that batts are fitted closely together at their ends and that all spaces around pipes or other intruding parts are packed solidly. Batts generally are applied on the inner face of the studs and their manufactured thickness allows a slight air space at the exterior surface. Installation in floor and ceiling is made in a similar manner by nailing to joists from the room side.

Rigid board insulation is manufactured from a variety of vegetable fibers compressed into a solid sheet, with the exposed surfaces covered with paper or asphalt. When used as sheathing the board is simply nailed onto the outer face of a stud wall and finish material is applied over it. Rigid board is also extensively used to insulate the foundation and floors in concrete slab construction and may be used as cavity-wall insulation in some brick construction. Installation in cavity walls is often difficult due to the numerous crossties used between the inner and outer wythes of wall, so pellet-type installation is a bit easier. When rigid board is used as roof insulation it is normally applied by the roofers and is set and coated with hot asphalt as a base for the exposed roof covering (Figure 7-2).

Pellet and granular insulation is usually installed after walls are covered each face, or when ceilings are in place. Because of the nature of the material it is virtually impossible to use this type of insulation in floors or roof construction or where there is a problem of containment. Various combinations of any of the insulation materials may add to the R value as necessary.

Insulation batt material is available in nominal widths of 16″ and 24″ and in thickness from 3½″ to 9″, paper or foil covered or without covering. The R value of fiberglass roll material will range from R-11 for 3½″ thickness to R-30 for 9″ thickness. Sheathing board is usually supplied in 48″ × 96″ sheets, while foamed glass roof insulation panels are available in thicknesses from ¼″ through 6″, in widths from 6″ to 48″, and in lengths from 24″ to 144″. Perlite (expanded volcanic glass) is one of the most used materials for pellets and will give an R value from R-4 upward, depending on the amount used and the type of containment. Blowing wool is supplied in bags of 25 lb each and requirements are easily figured as follows. For 1000 sq ft of R-11 insulation, use 11 bags with a coverage of not more than 90 sq ft at 4¾″ per bag. Similarly, a requirement for R-19 requires 19 bags; R-22 requires 22 bags; and so on. To estimate the quantities of insulation, simply determine the total gross area excepting openings over 50 sq ft, and purchase the proper thickness material for the R value required. If lamination or different layers are required, treat each layer separately.

Fiberglass batts are probably the most commonly used type of residential insulation and are normally installed by carpenter labor.

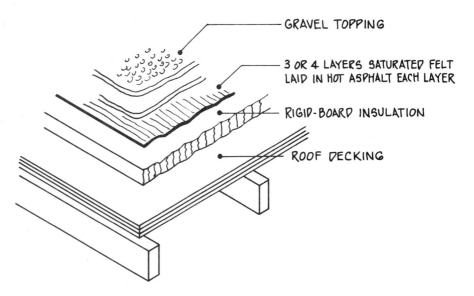

GRAVEL TOPPING

3 OR 4 LAYERS SATURATED FELT LAID IN HOT ASPHALT EACH LAYER

RIGID-BOARD INSULATION

ROOF DECKING

**Figure 7-2** Rigid-board roof insulation.

As stated before, the insulation R value has a direct relationship to the thickness of the batts. Thickness also determines to some extent the labor required to install this insulation. Batts normally 3½" thick are R-11 and a carpenter should be able to install about 1200 to 1500 sq ft per day at an in-place cost of about $0.40 per square foot. Batts of 6", R-19 value, cost about $0.55 per square foot in place. Rigid board 1" thick, R-4.3, is installed by one carpenter at about 700 sq ft per day at a cost of about $0.60 per square foot. Insulated foil-faced sheathing board costs about the same amount. Roof deck boards of fiberglass, 48" × 96" × 1⁵⁄₁₆" thick, R-5.3, are installed as part of the roofing at a rate of about 750 sq ft per day at a cost of $0.75 per square foot. Vermiculite or perlite pellets used for masonry fill insulation can be poured into core spaces of concrete masonry and cavity brick walls at a rate of about 1000 sq ft of wall per day and costs about $0.75 to $0.80 per square foot. The same material may also be figured for brick cavity walls at about 30 cu ft of material per hour at a cost of $2.30 per cubic foot. Poured insulation of fiberglass wool or mineral wool, R-4 per inch of thickness, installs at about 20 cu ft per hour at a cost of about $1.40 per square foot 1 inch thick.

## 7.2 ACOUSTIC TREATMENT

Acoustics is the treatment of sound, either to absorb it, as is most common, or to increase its density or direction, as may be required in an auditorium. The material used for *acoustic* control may be plaster in a number of forms, drywall with applied acoustic boards or tiles, suspended ceilings, and various wall treatments. In addition to materials used for construction, various acoustic properties are present in rugs and carpets, drapes, furniture, and even in the people who will occupy the room. Acoustic values or sound transfer is calculated by a coefficient of sound frequencies at 250, 500, 1000, and 2000 hertz (cycles per second) and is called the Noise Reduction Coefficient (NRC). This coefficient is represented by a single number to calculate the amount of sound-absorbing material required.

The NRC differs with each material, so another factor has been devised to simplify the selection of sound materials or combinations of materials. Sound Transmission Class (STC) is a number rating for airborne sound, representing the transmission loss of a wall or floor at all frequencies. Figure 7–3 shows a few methods for the control of sound transmission of different wall constructions and their approximate STC values.

In addition to the NRC and STC in sound control there are at least two other ratings involved: impact sound and transfer of sound through openings. The Impact Sound Pressure Level (ISPL) is the amount of sound transmitted through the solid construction and is most applicable in design and construction of second-story floors, particularly in apartment projects. Several methods used to reduce sound transfer through floors are shown in Figure 7–4 and are discussed in further detail in the text.

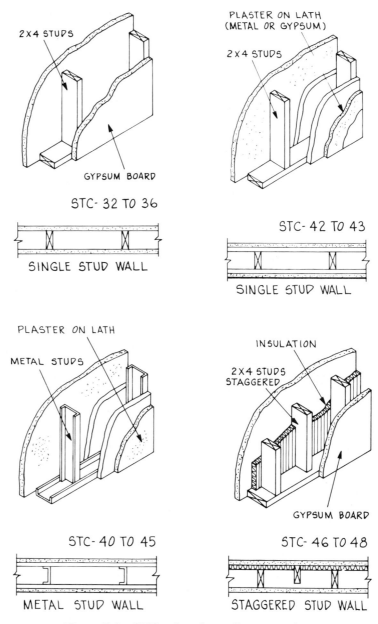

**Figure 7-3** STC values for wall construction.

## 7.3 ESTIMATING COSTS

For estimating purposes sound control methods may be divided into three major types: construction utilizing spaced studs and/or various layers of wallboard, absorbing material applied over other wall or ceiling surfaces, and several types of suspended ceilings. Transfer of noise may be by plumbing systems, improperly installed framing or finishing material, or by passage through cracks, holes, or uncaulked openings. Although it is definitely desirable to reduce sound, it is also possible to overdo it to the extent that a room may be "dead," with little or no sound reflected from walls, ceiling, or floor. In some cases it is almost impossible to control sound at windows or other openings if the outside level from traffic is high.

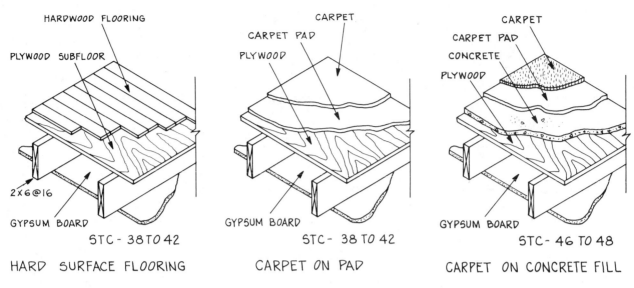

HARDWOOD FLOORING
PLYWOOD SUBFLOOR
2×6@16
GYPSUM BOARD
STC- 38 TO 42
HARD SURFACE FLOORING

CARPET
CARPET PAD
PLYWOOD
GYPSUM BOARD
STC- 38 TO 42
CARPET ON PAD

CARPET
CARPET PAD
CONCRETE
PLYWOOD
GYPSUM BOARD
STC- 46 TO 48
CARPET ON CONCRETE FILL

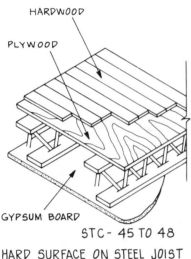

HARDWOOD
PLYWOOD
GYPSUM BOARD
STC- 45 TO 48
HARD SURFACE ON STEEL JOIST

**Figure 7-4** Sound transfer reduction through floors.

Interior partition walls installed with a complete floor-to-ceiling space between studs is often used in residential and apartment construction. This method is illustrated in Figure 7–3. Even if the walls are separated with an air space, the floor and ceiling construction may be continuous, thus transferring sound across the air space at top and bottom. Estimate this condition as two separate stud walls with the additional drywall or plaster finish figured separately. If additional layers of gypsum board are applied to deaden sound, estimate the wall at about $3 per square foot using two layers of ⅝″ gypsum board at a rate of 200 sq ft per day. If the space between the separated walls is filled with a fiberglass batt, add about $0.40 per square foot, one side. Unfortunately, even this wall system may not completely control noise from elsewhere in the building. Continuous pipe for water, ducts for air conditioning, and similar connected systems also conduct noise, and in many commercial systems these pipes or ducts are either fitted with separating insulators or ducts are lined with acoustic material to absorb the noise. For estimating interior duct lining, use about $2.50 per square foot of duct for lining and about twice that for labor, or figure labor at about 100 to 125 sq ft per day.

The ceiling is the home of most of the sound-absorbing material that may be nailed, screwed, stapled, or cemented to a backing. In most installations a plaster or drywall base surface is installed first, so estimate that material separately before the sound-absorbing material. If the estimator is careful and does not forget, the ceiling area taken off for the base material may be reused here to estimate the amount of sound-absorbing material. Most ceiling tiles are 12″ × 12″ and either ½″ or ¾″ thick, fiberglass or vegetable fiber, tongue-and-groove or flush edges, perforated or fissured faced, factory finished. Average insulation value is about 0.70 and the NRC is usually in the range of 0.40 to 0.55 with a light reflection about 75%.

Fire-retardant units are same size and thickness, NRC from 0.60 to 0.70, STC from about 35 to 44, and fire resistant 2 or 3 hr. Cost of installation for 12″ × 12″ ceiling tiles, ¾″ thick, mineral board, cemented to plaster or wallboard base, is about $1.25 per square foot and one carpenter should be able to install about 50 sq ft per hour. Wood fiber tiles 12″ × 12″, ¾″ thick, cost about 25% less in place. Fire resistance is now becoming a major consideration in the use of some materials, especially fiberboard and similar cellulose products, which are extensively used for insulation and acoustics. Mineral fire-rated tiles may be substituted for fiberboard at a cost of about $1.40 to $1.50 per square foot or "fireproofed" wood fiber board may be installed for about $0.85 per square foot in place. Primer may be necessary for some types of gypsum backing at a cost of about $0.25 to $0.40 per 100 ft, and installation cement averages about $10 per 1000 sq ft. Cement is used in small dabs at each corner of each tile so it is not a continuously spread material. If ceilings are rough or at slight differences in height, or are old ceilings being re-tiled, 1″ × 3″ furring at 12″ o.c. may have to be installed for a tile base at a cost of about $0.50 per square foot at a rate of about 50 sq ft per hour.

## 7.4 SUSPENDED CEILINGS

Another form of ceiling acoustical installation is by suspending a metal framework or grid, which in turn retains "lay-in" sound-absorbing panels. Although this system is usually used in commercial projects, it may also be used in residential work. The suspended ceiling grid is generally of aluminum "tee" sections interlocking at 12″ or 24″ intervals and hung from the roof or old ceiling at 48″ o.c. with #10 wires. This type of grid requires one carpenter about 1 hr per 100 sq ft of grid area and costs about $0.60 per square foot. If a grid system of Z bars or splines that will provide a finished flush surface is used, the cost is about $0.75 per square foot. With wood fiber "lay-in" panels, complete with suspended grid, the average in-place cost is about $1.75 per square foot at about 300 to 350 sq ft per day.

Commercial suspended ceilings often include lighting fixtures, which replace occasional acoustic panels. These ceilings may also include air-conditioning units. This type of suspended ceiling usually has enough space between the roof or ceiling and the suspended work to accommodate the various ducts, dampers, electrical conduit, and lighting boxes required. In some industrial installations the grid is strong enough and the plenum above it is deep enough to allow walkways or installation of minor pumps or fans. One of the major problems for the installation of this type of industrial suspended ceiling is the question of which union has jurisdiction: carpenters installing the grid, electricians installing the electrical fixtures, or "tin-knockers" installing the ducts and air conditioning. Fortunately, these are problems that do not concern the average residential work.

## PROBLEMS

**Sample problem:** Refer to the sample problem in Chapter 5. Calculate the amount of R-11 batt insulation in walls and the amount of R-19 insulation in the roof. What is a reasonable cost for this material installed?

| | | | |
|---|---|---|---|
| Rear wall: | 12.0 × 8.0 | 96 sq ft | Roof: 9.0 × 12.0 = 108 sq ft |
| Front wall: | 12.0 × 9.0 | 108 sq ft | Cost/sq ft  $ 0.55 |
| Two ends: | 8.0 × 8.0 | 64 sq ft | $ 59.40 |
| | | 64 sq ft | |
| End triangles: | 8.0 × 1.0 | 8 sq ft | |
| | | 340 | |

Less:                                          340 sq ft
  Door:     3.0 × 7.0  21.0 sq ft          − 35 sq ft
  Window: 3.0 × 4.5  13.5 sq ft            305 sq ft × $0.40/sq ft   $122.00
                          34.5 sq ft              Total cost              $181.40

**7-1.** If 1″ thick R-4.3 insulating sheathing board is applied on all exterior wall surfaces of the building described in the sample problem, what will be the total cost of the material, and how long will it take to install?

**7-2.** What is the amount of gypsum board required to sound-deaden the wall between garage and residence if the wall is 25′6″ long, 8′ to plate, and has a gable roof with a pitch of 2½ in 12? Use two layers of ⅝″ thick gypsum board each side installed over spaced wood studs.

**7-3.** Estimate the quantity of R-11 batts required to insulate the exterior walls and the R-19 batts in the ceiling of the house plan included in this book (see pages 194-197). Insulate the garage-residence wall but *not* the exterior garage walls.

# chapter
# 8

# Doors, Windows, and Glass

Almost any project, except possibly open sheds or warehouses, has a requirement for doors, windows, and glass in one form or another. In the case of large commercial or industrial projects, the material could take a wide variety of forms, from normal "people" doors for access, safety, or simply to close an opening, to larger hangar doors with single windows, to entire curtain walls of metal and glass. Since this text is not concerned primarily with large commercial or industrial projects, we will again describe mostly residential doors, windows, and glass in their smaller, more readily available, and better known forms. These types may be steel, aluminum, or wood for doors or windows, with most of the more standard forms of glass. Glass exceptions probably include the various special tinted, mirrored, or bullet-proof types, but include sheet glass, polished plate or float glass, obscure glass, and tempered glass.

In estimating items in this category the frames required and the operating hardware (normally known as "finishing hardware") need to be calculated in addition to the required doors, windows, or glass items. These again may be of the same materials as noted before or may be supplied as a part of the door or window assembly. Glass is usually a separate item, cut to the required size and shape, and installed in individual pieces. Mirrors are also made of glass but may be a part of the regular glazing if unfinished, may be purchased as framed items, or may be a part of a fabricated unit, such as medicine cabinets or certain other cabinetwork.

## 8.1 DOORS

Most residential doors are manufactured of wood. These can be classified into three principal groups: hollow core, solid core, and solid wood. The first two are built with solid wood stiles and rails and with solid "lock blocks" at possible lock locations. The interior spaces are then filled in with either a solid layer of wood strips glued together or with a grid or more open arrangement of filler material. In a few cases the interior spaces may be filled with gypsum or other mineral material to provide a fire-resistant unit. Entrance doors, Dutch doors, and a variety of other types are usually manufactured from solid lumber in a great number of designs. All of these doors are available as stock items from manufacturers and in a number of face veneers. Location, size, and number of units required are shown on the drawings and the type and face veneer is normally indicated in the specifications. In estimating work each type should be listed separately, by size, type, and surface material (Figure 8–1).

## DOOR SCHEDULE

| MARK | SIZE | TYPE | DESCRIPTION | FINISH |
|------|------|------|-------------|--------|
| ① | 3'-0" x 6'-8" x 1¾" | A | MAHOGANY | STAIN VARNISH |
| ② | 2'-8" x 6'-8" x 1⅜" | A | D. FIR | PAINT |
| ③ | 2'-8" x 6'-8" x 1¾" | B | D. FIR | PAINT |
| ④ | 2'-8" x 6'-8" x 1⅜" | C | PINE | PAINT |
| ⑤ | 2'-0" x 6'-8" x 1⅜" | C | PINE | PAINT |
| | | | | |
| | | | | |

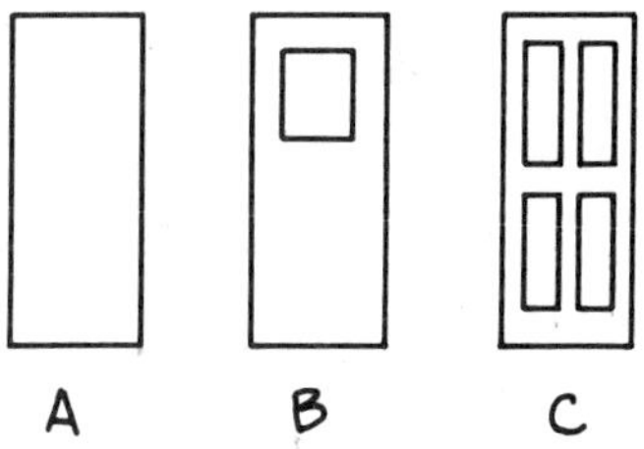

**Figure 8-1**  Typical door schedule.

Face veneer, the exposed surfaces on either side of the door, may be had in a number of common species such as Douglas fir, pine, several grades of birch or maple, or with processed hardboard. More exotic surfaces include hardwoods such as ash, oak, pecan, teak, or walnut. These are also available in a number of designs, such as "book-matched," "quarter-matched," "sequence matched," or other configurations utilizing the grain of the wood in a pattern. See Figure 8–2 for some typical patterns. Rails and stiles may be the same species as the face veneer or of a cheaper wood. Solid lumber doors are also available in Douglas fir, pine, birch, or any of the hardwoods. Door frames of wood again are usually of Douglas fir, pine, or similar softwoods when they will be stained or painted, but may be of hardwood to match hardwood door faces for clear finishes.

In many apartment uses the doors may be fabricated of two sheets of steel welded into a homogeneous unit over a metal grid or over a mineral filler blank. These doors customarily are supplied

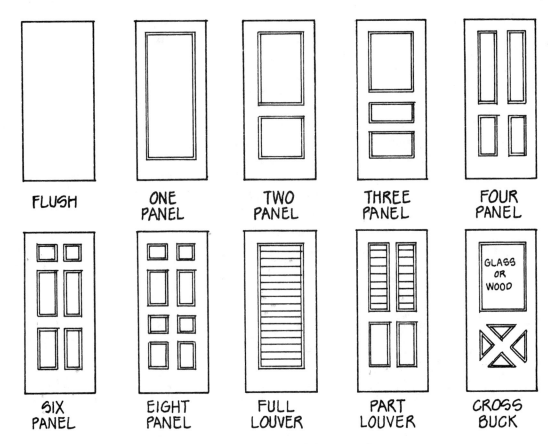

**Figure 8-2** Door faces.

complete with metal frames, and the entire unit is prime finished for future painting. When considered as "fire doors" they will be certified and tagged as being able to withstand a continuous fire exposure for periods from ¾ to 3 hr. Both the door and the frames will also be machined to accommodate the required door butts (hinges), locks, and other finishing hardware. When used in continuously damp locations, metal doors should be galvanized after fabrication to reduce the danger of rusting. Fire doors are rated in accordance with the requirements of Underwriters Laboratories (UL) or Factory Mutual (FM) standards and are classed as follows:

- A label: 3 hr
- B label: 1½ hr
- C label: ¾ hr
- D label: 1½ hr
- E label: ¾ hr

Required ratings for certain locations are specified by the fire codes and these should be checked if necessary. Fire doors and frames will cost about $100 for a 3'-0″ × 7'-0″ door with B or D label and require about 1 hr for installation.

Most "person doors" are manufactured in two principal

heights, 6'-8" or 7'-0", and in widths of 2'-0", 2'-4", 2'-6", 2'-8", and 3'-0". Thickness is usually determined by installed location using hollow-core doors or solid-core doors 1⅜" thick for interior use and solid-core doors 1¾" thick for exterior use. Special designs of solid wood doors for entrances may be as thick as 2". Wood frames are normally ¾" thick for lighter construction and 1⅛ to 1½" for entrance locations or heavy-duty installations. Most common sizes are 3'-0" × 7'-0" for entrance doors, 2'-0" × 6'-8" for bathrooms, and 2'-8" × 6'-8" for other interior doors. Movement of furniture through a door opening is one of the major considerations that determines door size.

## 8.2 ESTIMATING DOORS

Solid-core doors with birch veneer faces, 3'-0" × 7'-0", will cost about $60 for the door alone plus approximately $50 for the frame. Interior hollow-core birch-faced doors will cost about $40. "Pocket"-type doors that slide into a frame or multifold doors run from about $50 to as high as $150, depending on size and configuration plus $30 to provide the frames. Hardboard-faced hollow-core doors cost about one-half the cost of birch doors. Walnut-faced or other hardwood doors usually cost about 75 to 100% more than birch-faced doors. Hollow metal doors 3'-0" × 7'-0", complete with metal frames, cost from $150 to $250, depending on quality, and sliding aluminum-framed doors, often used for living room or patio access, cost about $600 in place, including hardware and glass.

Cost for the finishing hardware (butts, locks, closers, etc.) for doors is covered in Sections 8.7 and 8.8. However, installation of doors, fitting of the hardware, and related work adds another $25 to $30 per unit, employing an installation crew of two carpenters. Prefit wood doors, hollow-core, birch-faced, are available for about $125, including frames, and require about 1½ hr for a two-man crew to install. Other doors include folding accordian, bypassing or bifolding for closets, fire-rated, glass panel, Dutch doors, and a great number of special types. In estimating doors simply count the different sizes and types shown on the drawings and list and calculate them separately. If you do not have an exact cost, use the next *higher* cost to ensure that your estimate is sufficient.

## 8.3 WINDOWS

Windows, like doors, may be obtained in wood, steel, or aluminum for residential use. Bronze and some other materials are also available but are seldom used for residential work. Size and type of windows indicated on the drawings are usually determined by the required light and ventilation necessary for the enclosed room as dictated by the applicable building code. Windows may be solid fixed, vertical sliding (double-hung), horizontal sliding, bypassing, casement, drop-vented, up-swinging awning, or various combinations (Figure 8–3). Steel or aluminum windows are supplied complete with metal frames and all required hardware. Wood windows

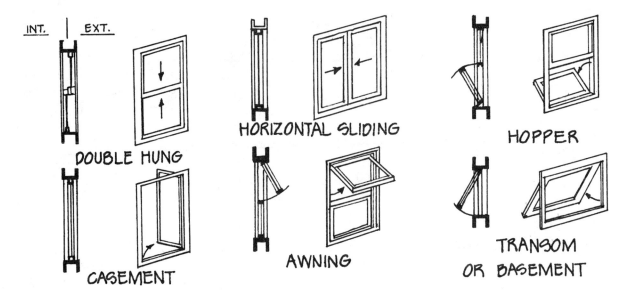

**Figure 8-3** Window types.

are also available in stock sizes and in most of the configurations that may be available in metal windows, complete with frames but not usually with operating hardware. Special sizes and types in both metal and wood may be fabricated individually. Glass is not usually a part of the windows except where specifically indicated. Hardware is supplied as a part of finishing hardware and installed with carpenter labor.

Metal windows have a factory finish. Steel windows are supplied with either a prime painted coat or galvanized; aluminum windows are anodized, either clear or possibly in color. Wood windows may be supplied without a factory finish, or prime painted or plastic coated. Concealed hardware is normally drop-forged, die-cast, or extruded from a base metal. Exposed *finishing hardware* is often bronze but may also be of other metal compatible with the window metal. Metal windows come complete with metal frames which have flanges on the edges for fastening to the wood structure, or may be embedded in masonry or concrete. When installed in wood frames the metal frames are made weathertight with waterproof paper and elastic caulking. Any weatherproofing required is done with the caulking material. Carpenters may install windows in wood frame construction, but in masonry and concrete construction the metal windows are usually installed as a part of the masonry or concrete work. Wood window frames in wood framing construction are shimmed for square and made weathertight with waterproof paper at jambs and with sheet metal flashings at head and sill. In the few cases where wood windows are installed in masonry or concrete construction the frames are provided with anchors but are installed in the proper location in a manner similar to that employed for metal windows.

## 8.4 ESTIMATING WINDOWS

Aluminum windows, complete with frames, glazed with grade B sheet glass, in sizes 3'-1" × 3'-2", cost about $35 for sliding type, $40 for single-hung vertical, $45 for projected vent type, and $85 for casement. Estimate installation costs about $35 to $45 per unit. Steel windows, glazed, sizes 2'-8" × 4'-6", cost about $50 for projected and $65 for double-hung. Installation costs are about the same as for aluminum windows. Wood windows, glazed, with frame and trims, size 2'-8" × 5'-0", cost about $85 for double-hung, $65 for 1'-10" × 3'-2" casement, $70 for 3'-6" × 2'-6" horizontal sliding, and $65 for 2'-10" × 2'-0" awning type. Installation costs about $35 to $50 per unit or about 2½ hr of carpenter labor for a 2'-6" × 4'-6" unit.

## 8.5 GLASS AND GLAZING

Glass in its many types (Table 8-1) is used in a great variety of locations in all kinds of construction and only a few special forms may not be used in residential projects. *"Sheet"* glass is formed by allowing the molten glass to flow over a flat plate; this process tends to include irregularities called "draw marks," small bubbles, and other defects that affect the use only in a very minor manner. When glass is formed in a thicker sheet than the ⅛" customary for sheet glass, but is not further finished, it is termed *heavy sheet* and usually

**Table 8-1**  Properties of glass

| Type | Thickness inch | mm | Max. available size | Approx. weight lb/sq ft | Kg/m² | Visible trans. % | Ave. reflect % | Solar trans. % | Rel. heat gains |
|---|---|---|---|---|---|---|---|---|---|
| Sheet |||||||||  |
| SS | 0.085 | 2.2 | 40 × 30 | 1.72 | 6 | 91 | 8 | 86 | 220 |
| DS | 0.115 | 2.9 | 60 × 80 | 1.64 | 8 | 90 | — | 86 | 215 |
| ³⁄₁₆ | 0.182 | 4.6 | 84 × 120 | 2.50 | 12 | 90 | — | 83 | — |
| Clear (float) |||||||||  |
| SS | ⅛ | 3.2 | 74 × 120 | 1.64 | 8 | 90 | — | 84 | 215 |
| DS | ³⁄₁₆ | 4.8 | 110 × 120 | 2.45 | 12 | 89 | — | 80 | 209 |
| ³⁄₁₆ | ¼ | 6.4 | 130 × 140 | 3.27 | 16 | 88 | — | 77 | 200 |
| Heavy duty |||||||||  |
| SS | ⅜ | 9.5 | 120 × 140 | 4.90 | 24 | 87 | — | 73 | 195 |
| DS | ½ | 13.0 | 120 × 300 | 6.54 | 32 | 86 | — | 70 | 190 |
| ³⁄₁₆ | ¾ | 19.0 | 180 × 300 | 9.81 | 48 | 83 | — | 63 | 180 |
| Heat absorb |||||||||  |
| SS | ⅛ | 3.2 | 74 × 120 | 1.64 | 8 | 83 | 7 | 63 | 179 |
| DS | ³⁄₁₆ | 4.8 | 110 × 120 | 2.45 | 12 | 79 | — | 55 | 161 |
| ³⁄₁₆ | ¼ | 6.4 | 120 × 192 | 3.27 | 16 | 75 | — | 47 | 135 |
| Grey |||||||||  |
| SS | ¼ | 6.4 | 120 × 192 | 3.27 | 16 | 42 | 5 | 47 | 150 |
| DS | ⅜ | 9.3 | 122 × 164 | 4.91 | 24 | 29 | — | 33 | 130 |
| ³⁄₁₆ | ½ | 13.0 | 120 × 130 | 6.55 | 32 | 20 | 4 | 24 | 113 |
| Bronze |||||||||  |
| SS | ¼ | 6.4 | 124 × 204 | 3.27 | 16 | 51 | 6 | 46 | 168 |
| DS | ⅜ | 9.5 | 122 × 264 | 4.91 | 24 | 35 | 5 | 31 | 156 |
| ³⁄₁₆ | ½ | 13.0 | 120 × 300 | 6.55 | 32 | 27 | — | 22 | 130 |

is about $\frac{7}{32}''$ thick. Polished plate glass ranges from $\frac{1}{4}''$ to as much as $2''$ thick and is machine polished parallel both sides. A newer method of providing a similar type of glass is by flowing the liquid glass material over a bed of molten mercury and is called *float* glass.

Special glass includes wire embedded in the molten glass material to supply break resistance in cases of fire or physical blow, "shatterproof" glass manufactured with a sheet of plastic laminated between two glass sheets, "tempered" glass with surfaces in tension and all edges sealed, colored glass in a number of colors for heat or energy control, one-way mirror glass which allows vision from one side but is a mirror on the opposite side, "bullet-proof," and a number of others. Mirrors are made of plate glass with a silver and copper backing applied, and may be supplied with metal or wood frames or without frames but with ground and polished edges. Insulating glass is formed by fabricating two sheets of glass with an air space between and with the edges sealed to provide a dead-air space between the glass face sheets. Glass that is "tempered" or has edges sealed should not be cut on-site, as this will probably destroy the entire unit when tension is released.

## 8.6 ESTIMATING GLASS

Sheet glass of grade B is used for glazing up to 16 sq ft of area and costs about $2.75 to $3.00 per square foot in place. Patterned "obscure" glass, $\frac{1}{8}''$ thick, used for bathroom windows and similar locations, costs about the same as clear glass. Plate glass $\frac{1}{4}''$ thick, clear, costs about $3.50 to $4.00 per square foot in place but approximately $6.50 per square foot for tempered glass. Insulating glass consisting of two sheets of $\frac{3}{16}''$ float glass fabricated to a total thickness of $\frac{5}{8}''$, clear, costs from $8.50 to $9.00 per square foot in place. Wire glass, $\frac{1}{4}''$ thick, clear or obscure, is about $4.00 per square foot in place. Mirrors, without frames but with edges ground and polished smooth, $\frac{1}{4}''$ polished plate, in sizes to 6 sq ft, cost approximately $4.25 per square foot.

Glass is often estimated by "united inches" (height plus width) and the crew size for installation by one man up to 100 sq in. and for two or three men beyond that area. For glass to $\frac{1}{4}''$ thick one man should be able to putty-set to 50 lites each 3 sq ft in area in 8 hr, but this production drops to 15 lites when the "united inches" reach 135 and requires two glazers. Insulated glass, tempered glass, and special glass installation costs about 100% mor than for $\frac{1}{4}''$ polished plate. Irregular shapes and long narrow lites also reduce installation by 30 to 50%, and access for installation may further reduce production by as much as 100% or more, depending on location of installation.

## 8.7 FINISHING HARDWARE

The term "finishing hardware" includes all hardware items that are exposed or that are not considered "rough hardware." In the latter category are nails, screws, bolts, lags, miscellaneous metal clips, and smaller parts normally used to fasten parts of the structure together.

**Table 8-2** Standard U. S. hardware finishes

| USP | Primed for painting | US15 | Dull nickel plated |
|-----|---------------------|------|--------------------|
| US1B | Bright Japan (black) | US19 | Sanded dull black |
| US2C | Cadmium plated | US25 | White bronze |
| US2G | Zinc coated | US26 | Chromium plated |
| US3 | Bright brass | US26D | Chromium plated, dull |
| US4 | Dull brass | US27 | Satin aluminum, lacquered |
| US9 | Bright bronze | US28 | Satin aluminum, anodized |
| US10 | Dull bronze | US32 | Stainless steel |
| US14 | Nickel plated | | |

Finishing hardware includes all the hardware items required to "finish" a project, especially doors, windows, cabinets, and similar items. Doors require butts (hinges), locks with keying, latch sets with knobs but no keying, pulls, kickplates, door closers, hold-open devices, and panic hardware. Cabinets require hinges (not termed "butts"), catches, locks, pulls, drawer glides, and other small parts.

Finishing hardware is priced by three factors: quality, finish, and use. Most manufacturers are able to supply a complete assortment of quality from least expensive tract housing material to luxury grade. Finishes for exposed parts may be aluminum, stainless steel, plated brass or bronze, black iron, or painted and are fairly uniform, in accordance with accepted finish designation standards (Table 8-2). Concealed parts in the least expensive quality may be stamped steel, die-cast metal, or partially finished machined parts; in superior quality parts may be cast, forged, or completely machined from solid metal. A few lower-priced items are now using plastics for some parts. The frequency of operations, interior or exterior location, material to which applied, and code requirements determine the third factor. These three factors combine to determine the cost of the finishing hardware and are indicated to the supplier in two major ways: by a piece-by-piece list of each item for each opening, door, window, or cabinet, or by an "allowance" estimated for the material required. In the first case each opening has a list identifying the butts, lock sets, and other items by a known manufacturer's stock number. This creates an undesirably long list subject to possible error in transposing stock numbers or by omitting some items. This difficulty may be relieved in part by identifying like openings by an identical symbol (door A, door B, etc.) and by specifying the required finishing hardware only by type of opening instead of listing every opening, even if identical. Unfortunately, this system is not of much help in residential work simply because there are not that many doors of the same type, except perhaps in apartment construction. A method often used for residential work is by allowance. Here an estimated amount of money is included in the contract figure for purchase of the finishing hardware without specific identification of the items. The owner, together with the architect, contractor, and hardware supplier, then select the required items from stock and adjust the material in accordance with the allowance agreed upon. Labor of installation should not be included in the allowance figure, as installation is part of the contractor's production costs.

## 8.8 ESTIMATING FINISHING HARDWARE

Many estimators arrive at finishing hardware costs by using a percentage of the total building cost. The R.S. Means Co., Inc., quotes "average percentage, total cost of job, minimum at 0.5% to a maxium of 3%, with "average distribution for materials at 85% and 15% for labor." This does not include hardware required in addition to butts and lock sets, so items such as door closers ($50 to $60 in place), panic devices ($200 to $500 in place), or other items usually considered with commercial or industrial construction must be estimated in addition where required.

Labor for installing finishing hardware is considered a carpenter item. Carpenters set door frames, fit the doors to the frames, mortice, cut, or otherwise prepare both frames and doors for the finishing hardware, and install the hardware required. Average time for trimming and fitting the door blank for a wood door 3'-0" × 7'-0" is about 2 hr plus another 2 hr for installation of frame and trim, or about $75 to $80 per door. Metal door frames and metal doors are morticed and drilled for finishing hardware by the manufacturer if the hardware templates are provided for this purpose.

## 8.9 RELATED ITEMS

Garage doors may be factory manufactured in a great number of types and sizes in both wood and metal and are usually supplied complete with hardware. This hardware may be the economical type, using coiled side springs and steel channels or angles attached to the door, or may be an overhead track system with the door on rollers. Garage doors of wood, size for a "double" or two-car garage 16'-0" × 7'-0" will cost $400 each in place. Metal doors, often horizontally sectional, of the same size will cost about $275 in place. Electric operators for remote control are approximately $275 additional.

Screen doors may be of metal or wood with metal or plastic mesh and normally do not require frames in addition to the door frames. Screen doors will cost about $15 to $35 in place. Screens for metal windows are often included as a part of the metal window assembly or may be estimated at $2.25 per square foot. Storm windows will average about $30 for a 4'-6" × 5'-3" opening.

## PROBLEMS

**Sample problem:** Estimate the cost of doors for an industrial building using ten 3'-0" × 7'-0" solid-core birch-faced doors and six hollow-core birch-faced doors, complete with frames and an allowance for finishing hardware of $56 per door.

|  | Material | Labor | Total |
|---|---|---|---|
| Doors: 10 solid core at $60 each | $ 600 | | |
| 6 hollow-core at $40 each | 240 | | |
| Frames: 16 at $40 each | 640 | | |
| Finishing hardware: 16 at $56 each | 896 | | |
| Labor for installation: 16 doors at $75 | | $1200 | |
| | $2376 | $1200 | $3576 |

**8-1.** Compare the material costs for 12 steel windows, each 2'-4" × 4'-2", prime painted against the same number and comparative size of clear-anodized aluminum windows.

**8-2.** Make a proper take-off for 50 lites of DSB (double-strength grade B) glass, each lite 8" × 10", plus two panes of ¼" polished plate glass each 4'-6" × 6'-8", plus two mirrors, edges ground and polished, each 24" × 36" × ¼".

**8-3.** Using the house plans on pages 194-197, estimate the total cost of the birch-faced doors, steel windows, and all glass.

chapter

# 9

# Floors

Floor construction in any structure forms a horizontal diaphragm that stiffens the building in addition to separating it vertically. Floors may be fabricated of almost any material, depending on their installation as either subfloors or finish floors, and may be installed on a solid bearing surface such as the earth or may be suspended in a number of ways. In residential work, especially for multistoried homes, there are usually combinations of earth-supported and framing-supported floors, and both subfloors and finish floors.

## 9.1 CONCRETE FLOORS

The most common earth-supported floor is of concrete installed over a gravel base or moisture barrier of plastic film or waterproof building paper. In a garage area and similar spaces the floor may be finished by smooth troweling and also serve as the finish floor. Where a concrete on-grade slab is used for the living portion of the building it is usually covered with resilient flooring, carpet, strip or parquet wood flooring, or even stone or masonry units. The concrete floor may be reinforced with deformed steel bars or welded-wire mesh to reduce cracking and is normally placed in a continuous operation with edge footings or concrete foundation portions. When concrete is used over a wood floor to reduce noise transfer or as a fire stop, there should always be a moisture barrier installed between the two materials. Temperature distribution materials such as welded-wire fabric or stucco mesh (chicken wire) will be useful when embedded in the concrete slab.

Concrete for on-grade installation should have a maximum heavy aggregate size not larger than ¾" in order to provide a smooth exposed surface, and should have a compressive strength of 2000 psi or more. Forms for edges below grades may not be required, allowing placement of concrete directly against the excavated earth surfaces, but edges above the grade will need some type of form to retain the concrete in required shape and position (see Chapter 4 regarding concrete forms). Cost for concrete is by the cubic yard (36″ × 36″ × 36″), but most slabs are only 4″ to 6″ thick, so careful calculation of required cubic yards should be made. This is probably done most easily as a combined figure when calculating concrete foundation requirements, but when indicated as a covering over wood subfloor or a steel deck should be taken off as the gross area divided by figure representing the slab thickness in feet.

Finish of concrete floors varies with the ultimate use. For exposed concrete floors in garages or similar areas the finish is a simple troweled finish giving reasonable smoothness but not "polished." For walks and similar areas the finish may be "broomed" or slightly rough to prevent slipping. Where resilient flooring is to be installed, however, the floor must be smooth and troweled "until the trowel emits a ring" when used. Special care should also be used in the selection of curing compounds for concrete to ensure that they are compatible with the adhesive that may be used to install resilient flooring.

## 9.2 RESILIENT FLOORING

Resilient flooring covers a wide range of materials, including asphalt tile, vinyl asbestos tile, vinyl sheet material, rubber, cork, linoleum, and other semiresilient materials. Almost without exception these materials are glued to the subfloor with a proper adhesive or factory-applied material. Most tiles are 9″ × 9″ or 12″ × 12″ and are available from 1/16″ to 1/4″ thick. Sheet material is supplied in rolls nominally 72″ wide and in various thicknesses, depending on the material. Pattern matching may be important at edges of rolls but is not of particular interest in tiles except perhaps in "marbleized" asphalt tile. In addition to the floor material a plastic or rubber base, 4″ or 6″ high, is usually required at all walls to trim out the floor (Figure 9-1). Finish schedules on the drawings should indicate the rooms that are to receive resilient flooring, while the specifications will indicate the quality, pattern, and probably the acceptable manufacturer.

Resilient flooring is calculated in square feet of installed area. Be sure to quickly check room sizes to determine if tiles will lay without excessive waste. Most tile laying requires a starting point at the center of the room in order to provide nearly identical edge tiles at wall lines, and this may mean cut tiles at all walls. Resilient base is calculated in lineal feet and may be supplied in units 48″ long or as a continuous roll. Disregard openings for doors, as the labor to trim around doors is often more than for straight lengths and the amount of material saved is insignificant. In some eastern and midwestern areas of the United States a layer of "deadening felt" underlayment is first cemented to a wood subfloor before resilient flooring is installed. Floor tile and base should include about 10% average waste.

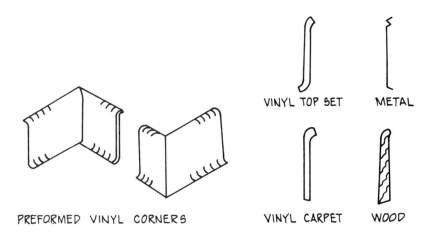

VINYL TOP SET    METAL

PREFORMED VINYL CORNERS    VINYL CARPET    WOOD

**Figure 9-1** Typical bases.

## 9.3 COST OF RESILIENT FLOORING

Asphalt tile is the old standby in resilient floor covering and is supplied in grade A (solid black, green, or brown) through grade E (white, yellow, light colors), with the C grade being the most commonly used. Tiles are individually installed after the area has been primed and a coating of adhesive applied and cost from $0.65 to $0.75 per square foot in place. Vinyl asbestos is a step upward in quality and $12'' \times 12'' \times \frac{1}{16}''$ tiles cost about $0.75 to $0.80 in place. Cork tiles are available in a variety of brown-tan patterns and average about $1.50 per square foot in place. Pure vinyl tiles are not generally very common in residential construction due to their high cost, up to $4.00 per square foot in place. Rubberized marbleized tiles are also not very commonly used and average about $1.80 per square foot in place. Sheet material obviously will have more waste, so multiple full widths of $72''$ may have to be calculated for material and costs about $1.50 for rubber to $2.50 for $0.093''$ thick vinyl. Coved rubber or vinyl base, $6''$ high in standard colors, averages about $0.95 to $1.00 per lineal foot in place, with premolded corner units at approximately $1.15 each in place. Adhesive at about $6.50 per gallon should be figured at 1 gal for each 200 sq ft of flooring surface. Labor for resilient flooring installation will average about 3 hr per 100 sq ft for most types, including roll material.

## 9.4 CARPET

Carpet is manufactured in such a variety of quality, pattern, color, and style that it is very difficult to estimate unless the cost per square yard or an allowance per square yard is known. Residential carpet is usually in the range $8 to $12 per square yard and requires padding of some kind beneath the carpet to give satisfactory results. The padding is usually glued to the subfloor, but the actual carpet may be glued solidly or stretched and pinned to tack strips at walls. Seams between rolls are sewn together and taped in good installation and the pattern is carefully matched. The latter requirement may cause some waste as will finished width of rooms that do not consider nominal 9 ft and 12 ft manufactured widths of materials.

Calculated carpet requirements are by square yards plus about

10% allowance for average waste. Padding beneath the carpet is calculated in the same manner as the carpet. Padding will cost from about $1.50 per square yard for jute to as much as $5.00 per square yard for heavy sponge rubber, in place. If rubber or vinyl top-set base is required, it is figured in lineal feet, the same as for resilient flooring installations.

## 9.5 WOOD FLOORING

Wood finish floors are usually installed over wood subfloors and may be "softwoods" such as fir or pine but are more commonly of a hardwood species such as maple, oak, pecan, teak or walnut. Strip flooring is available in a number of widths, thicknesses, and grades, the most common being $^{25}/_{32}$" × 2¼" with quality from clear select to #1 common. Almost without exception strip flooring is manufactured edge matched and end matched. This means that a tongue-and-grooved joint is possible at both edges and ends. Wood floors are installed over a layer of deadening felt or building paper and the strips are edge-nailed or screwed at the tongue side to conceal all fastenings when the next board is installed. Face nailing is unprofessional and presents a poor appearance. An expansion space of approximately ½" is allowed at walls and is usually covered with a wooden baseboard attached to the wall. Wood floors installed on concrete subfloors require a good waterproofing of the concrete and in known damp areas should be avoided or installed on pressure-treated wood sleepers fastened to the concrete, thus allowing a dead-air space between concrete surfaces and the finish wood floor. Strip floors are seldom installed with prefinished material so will have to be sanded and finished as an additional cost.

Parquet floors are factory manufactured into separate sheets with identical pattern and with edges tongue-and-grooved or cut for metal splines. The individual face pieces are either grooved together or fastened to a base material without nailing. Thickness runs from $^{5}/_{16}$" to ¾". There are about 10 standard patterns available from most manufacturers, using about 10 hardwoods (Figure 9–2). Wood grain, unit arrangement, natural or stained color, unfinished or factory-finished exposed surfaces, and species of wood determine the unit cost. Even though adequate subfloor is not detailed on the drawings, the estimator should include ¾" plywood or 1" nominal boards topped with ⅜" underlayment in his carpentry material and labor take-off. An inadequate subfloor will continue to give trouble long after the installation has been completed. Parquet is usually installed in a heavy mastic but may also have edge nailing if necessary.

Plank floors are somewhere between strip flooring and parquet in appearance. Whereas strip flooring is usually in narrow pieces, plank flooring material varies from 1½" to 8" in width and is ¾" thick. Some patterns use alternating widths, false peg representation, bevel edges, as well as herringbone, or other face configurations. Plank flooring is installed in the same manner as strip flooring and may be obtained unfinished or factory finished.

| PATTERN | SIZE | THICKNESS | WOOD | GRADE | WEIGHT LB. 1000 ' |
|---|---|---|---|---|---|
|  STANDARD | 19 × 19 | 5/16 | CHERRY, MAPLE, OAK, CEDAR, PECAN, WALNUT, TEAK, PANGA | PREMIUM SELECT | 1250 |
|  HADDON HALL | 13¼ × 13¼ | 5/16 | TEAK, W. OAK, R. OAK, PANGA, WALNUT | SELECT BETTER | 1250 |
|  MONTICELLO | 10 × 10 13¼ × 13¼ | 5/16 | TEAK, W. OAK, R. OAK, PANGA, WALNUT | SELECT BETTER | 1250 |
|  CANTERBURY | 13¼ × 13¼ | 5/16 | TEAK, W. OAK, R. OAK, PANGA, WALNUT | SELECT BETTER | 1250 |
|  SAXONY | 19 × 19 | 5/16 | TEAK, W. OAK, R. OAK, PANGA, WALNUT | SELECT BETTER | 1250 |
| 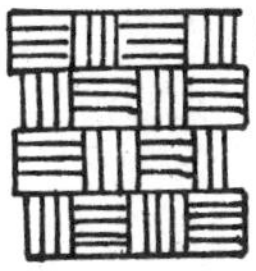 BASKETWEAVE | 15½ × 19 | 5/16 | TEAK, W. OAK, R. OAK, PANGA, WALNUT | SELECT BETTER | 1250 |
| 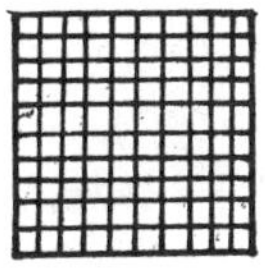 DOMINO | 18 × 18 19 × 19 | 5/16 | TEAK, W. OAK, R. OAK, PANGA, WALNUT | PREMIUM | 1250 |
|  HERRINGBONE | 14⅛ × 18⅛ | 5/16 | TEAK, W. OAK, R. OAK, PANGA, WALNUT | SELECT BETTER | 1250 |
| INDIVIDUAL PLANK | 3" TO 8" | ¾ | TEAK, W. OAK, R. OAK | SELECT | 2700 |
| LAMINATED BLOCKS | 6¾ × 6¾ 9 × 9 | ½ | OAK, PECAN | SELECT | |

**Figure 9-2** Parquet patterns.

## 9.6 COSTS OF WOOD FLOORING

Strip flooring is estimated in number of square feet of area and should include 20% for narrow stock and 40% for material more than 2″ wide. About 50 lb of flooring nails is required per 1000 sq ft of flooring. Labor for installing 1000 sq ft of strip flooring is about 20 to 25 hr at an average of 300 sq ft per 8 hr. Machine sanding of installed unfinished wood strip flooring requires about 1 hr of labor per 100 sq ft. Sanding and finishing costs about $0.85 to $1.00 per square foot.

Strip flooring of maple or oak, nominally $^{25}/_{32}$″ × 2¼″ in size, nailed in place over wood subfloor, will cost about $2.50 to $3.00 per square foot. Parquet flooring in separate units installed unfinished will cost about $3 in a simple oak pattern to as much as $10 per square foot in teak or walnut. Plank floors installed unfinished will average about $4.00 per square foot at 150 sq ft per day. Pattern and wood species are the two determining factors in parquet flooring, so the estimator may want to check carefully with a supplier before making a firm commitment.

## 9.7 MISCELLANEOUS FLOORING

Linoleum is rapidly becoming an obsolete flooring material but may still be used for some installations. Linoleum is a printed pattern in color, imposed on a foundation of burlap covered with a composition of various wood granules, linseed oil, and adhesives. Material is usually available in 6-ft-wide rolls, $\frac{1}{16}$″ to $\frac{1}{8}$″ thick. Linoleum is laid over a felt underlayment glued to wood subfloor with linoleum paste. Pattern may determine the number of strips required, but the number of seams should be minimized as much as possible. The cost of material varies with pattern and quality, but a linoleum layer should be able to install about 50 to 60 sq yd of deadening felt per 8 hr and 30 to 35 sq yd of linoleum in 8 hr.

Flat stone (flagstone), slate, brick, and possibly terrazzo are also used for residential flooring materials. These products are special in the fact that they are usually used only in entries or other locations where durability, not resiliency, is of prime importance. These are all usually installed over concrete bases with cement mortar. Stone, slate, and brick are included in discussions in Chapter 4, and tile floors are included in Chapter 12. Terrazzo is a "wet material" in that it is installed as a mortar mix material, then ground and polished after it has set, and costs about $4.00 to $5.50 per square foot in place. A terrazzo worker and one helper should be able to install about 125 sq ft of terrazzo per day.

## PROBLEMS

**Sample problem:** Estimate the cost of installing 9″ × 9″ × ⅛″ asphalt tile, grade C, in a room 9′-6″ × 13′-4″, 4″ top-set base, two doors each 2′-8″ wide.

Floor area: 9.5 × 13.33 =    126.64 sq ft at $.75           $ 95.25
Base: 9.5 × 13.33 × 2  =     45.66 lin ft
   Less: 2 doors at 2.66     −  5.33 lin ft
                        40.33 lin ft of base at $1.00     41.00
Adhesive 1 gal for 126 sq ft plus base at $6.50           6.50
Time required: about 6 hr                          $142.75
                                        SAY $150.00

**9-1.** What quantity of sheet vinyl would be required for the sample problem above using 72 ″ wide sheet material with a 4 ″ coved base at all walls? Is there any advantage in direction of installation of the short way of room over the long way?

**9-2.** What will be the cost for installing random-width edge-matched end-matched oak strip flooring in the sample problem above?

**9-3.** Estimate the amount of carpet that may be required for the living room in the house plans on pages 194-197. Rolls will be supplied 12′-0″ nominal width and to 100 lin yd long.

# chapter
# 10

# Walls
# and
# Ceilings

Walls for construction come in all shapes and sizes and are formed with almost any material. In residential work there are fortunately not as many options as in commercial work, so we may only be concerned with two major classifications: "wet walls" formed with concrete, plaster, or stucco, in which water is a prime ingredient, and "dry walls" of gypsum board, lumber, or plywood, in which water is not required. Either of these two types or combinations of both may be used to form walls. In this same consideration we have to be aware that ceilings are formed in somewhat the same way as walls and may be made of any of the materials mentioned.

Exterior walls give the project shape, most often support all or a major part of floor, ceiling, and roof framing, and when properly braced, strengthen the structure to resist earthquake and wind forces acting laterally. Interior walls may also help support floor, ceiling, and roof framing if so designed, generally acting as partitions or separators to form various rooms, and except in a relatively minor way, do not add to lateral bracing. Residential framing of walls is usually accomplished by a system of vertical studs and horizontal plates braced diagonally at corners and intermediate points. Interior partitions are installed in a similar manner. Estimating of wood rough framing is covered in Chapter 5.

Concrete or masonry walls may be considered *wet walls* since the concrete, mortar, and grout require water to develop a proper mixture; however, these materials are considered in Chapter 4. We are therefore interested principally in the material used to cover the structural framing and to enclose and divide the building into its various rooms and spaces.

## 10.1 METAL WALL FRAMING

Although most walls in residential construction are framed with wood, there is increasing interest in and use of metal. Metal wall framing systems have been common in commercial projects and provide better fire resistance than wood. Metal framed walls consist of metal studs connected at top and bottom with a metal "track" or U-shaped channel. This frame is then covered with expanded metal lath and plaster or may be covered with gypsum drywall or other dry material. Bracing within the metal frame is usually fabricated of small metal channels installed at a diagonal and secured to the studs and top and bottom "plates." The vertical studs are secured to the plates with sheet metal screws, tie wire, or in some instances by spot welding. Studs are available from $1\frac{5}{8}$" wide to 6" wide and spaced at 16" o.c. will give adequate support for walls from 8'-0" to 20'-0" high.

Metal lath is available in several types, as shown in Figure 10–1. Most common is "expanded" metal lath made by perforating a sheet with parallel slits, then stretching the sheets to upset the solid strips to form a diamond-shaped open pattern. Metal lath is generally specified as 2.5 or 3.4 (weight of 1 sq yd), black dip-painted finish or galvanized. Metal lath is tied to the studs with 18-gauge annealed wire at about 6" o.c. or nailed to wood studs with large-head roofing nails. Asphalt-saturated building paper (felt), normally 15 lb per 100 sq ft, is often applied behind the mesh to form a barrier for plaster and to reduce plaster waste. Edge and end laps of lath sheets should be at least 2" and should be firmly tied together. Where additional strength is required, various forms of ribbed-lath which have a solid portion strengtening the sheet lengthwise may be used. Metal lath is not regularly used for exterior plaster (stucco) except for the soffits of eave overhangs and similar locations. Exterior lathing, or stucco mesh, is discussed later in this section.

Metal lath work includes a number of accessory parts. Corners are almost always reinforced with a "corner bead" unit for outside corners and a 6"-wide strip of metal lath for inside corners. Expansion joints are W-shaped metal with wings each side of the mesh. Edge beads are of a number of different shapes, but all are intended to finish out the plaster smoothly and to act as "grounds" for the plaster installation. Metal bases of several shapes and sizes are also available. Metal access doors are also a part of lathing and plastering and usually are supplied complete with door, frames, and attachment wings where access to attic or similar spaces is required.

Metal wall framing is normally done by "lathers," although lathing work and the subsequent plastering may be done by one contractor. Walls that are not supporting a floor or roof load (termed "nonbearing") are usually formed with steel studs of 25-gauge metal, $3\frac{1}{4}$" or $3\frac{5}{8}$" wide, spaced 16" o.c., and require about 2 hr of labor for one installer for each 100 sq ft of wall area. In-place costs for metal studs vary from about $0.60 per square foot of wall to about $0.90, depending on size and spacing. Diamond pattern 3.4-lb metal lath wired to steel studs averages about $3.50 to $3.70 per square yard in place installed at a rate of 60 to 80 sq yd per day.

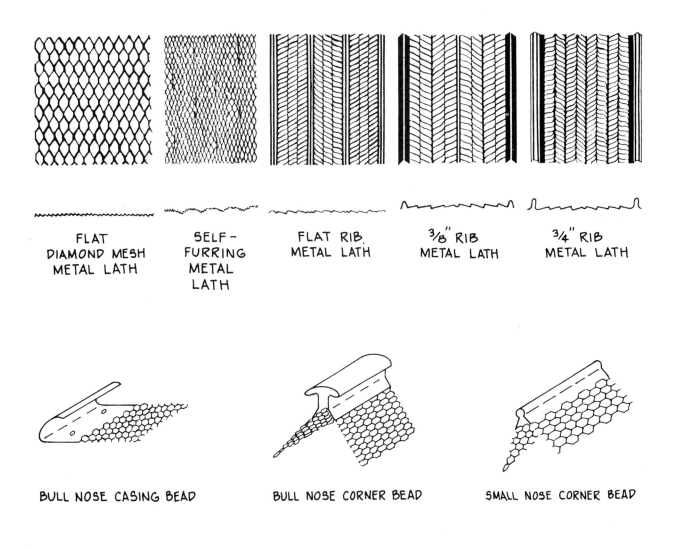

FLAT DIAMOND MESH METAL LATH

SELF-FURRING METAL LATH

FLAT RIB METAL LATH

3/8" RIB METAL LATH

3/4" RIB METAL LATH

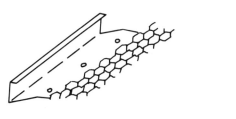

BULL NOSE CASING BEAD

BULL NOSE CORNER BEAD

SMALL NOSE CORNER BEAD

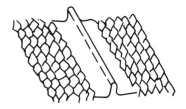

SQUARE CASING BEAD

BASE OR PARTING SCREED

**Figure 10-1** Types of metal lath and trim. (Courtesy of California Lathing and Plastering Contractors Association.)

Ribbed lath costs about $0.25 per square yard more in place.

A considerable amount of accessory metal parts are necessary for plaster installations over metal studs. Corner beads for either "inside" or "outside" corners are wire-tied in place at a rate of about 250 lin ft per day and cost about $0.75 per lineal foot in place. Expansion joints are necessary for large areas and cost $1.25 per lineal foot in place. Casing beads at openings are used as screeds for plaster and are comparable in cost and placement to corner beads.

"Furring channels" are small cold-rolled steel members, usually 16 gauge, in sizes from ¾" to 2" deep and are used to brace studs, to form around columns, and to reinforce other locations. Installation in walls varies greatly but may be estimated at about $0.75 per square foot of wall area.

Exterior walls to be covered with plaster or stucco also require a metal fabric base material known as "stucco mesh." Fabricated from 17-gauge wire in a diamond pattern with holes about 1½" maximum length and weighing 1.8 lb per square yard, it is commonly termed "chicken wire." When used as a base material for exterior cement plaster it is installed over 15-lb waterproof building paper and held in place with double-head furring nails which allow the wet plaster to more-or-less enclose the wire in the middle of the plaster. Stucco mesh applied over wood framing costs about $8.00 per 100 sq ft for the paper and about $2.10 to $2.25 per square foot for mesh. Estimate gross square feet of area disregarding openings smaller than 20 sq ft. Allow 10% for waste and edge lapping. "Outside" corners should be added assuming about 18" on each wall or a sheet 18" wide by the height of the corner.

## 10.2 CEILINGS

Ceilings, similar to walls, may be framed with wood or metal, but in most residential work are framed of wood. Suspended ceilings are usually used when the rough framing is to be concealed, a plenum is necessary for installation of heating ducts or electrical work, or for acoustical reasons. Most suspended ceiling work is framed with cold-rolled channels or channel studs installed horizontally and suspended from the rough framing with #10 wire at 48" o.c. each way. Metal lath may be wired to this framing or acoustical material may be applied directly to the frame or surface material (Figure 10–2). Installation of ¾" channels for suspended ceilings is about 150 square feet per day and costs about $1.25 per square foot. Cost and installation of carrier for members across these channels should be about the same giving a total of $2.50 for the metal framing. Metal lath is wired in much the same way as for walls at about 50 to 60 square yards per day at an approximate cost of $4.25 per square yard. Ceilings generally have few openings so the entire gross area should be used plus about 10% for waste and lapping requirements.

## 10.3 OTHER PLASTER BASES

Metal lath or stucco mesh are obviously not the only materials that may be used as a base for plaster. For many years wood lath, each 1½" × 48" × ⅜" thick, were installed over wood studs. The lath were rough-sawn and nailed to studs about ⅜" apart and ends were staggered to break any joint pattern. In some remodel work wood laths are still used to match the existing work but require too much time for economical application today.

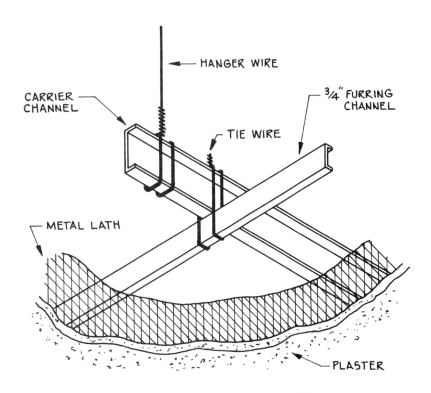

**Figure 10-2** Suspended plaster ceiling.

Gypsum lath have almost completely replaced wood lath. Available in nominal sizes of 16″ wide by 48″ long and ⅜″ to ⅝″ thick, lath are sheets of dry gypsum covered each side with absorbent paper and are obtainable in solid surface units or with a uniform system of punched holes. In the early stages of gypsum lath production all sheets were supplied with punched holes which allowed the plaster to be forced through the hole and added an advantageous holding factor to retain the plaster in position. With the introductions of smooth-faced gypsum lath the facing paper was manufactured with greater suction ability and the holes were eliminated. Insulating gypsum lath has a foil surface one side and this side should be applied opposite to the plaster surface. Gypsum lath is installed with tight, but not forced, joints and wide joints or holes should be reinforced with metal lath. Corners and eges are also reinforced with metal lath or standard corner beads manufactured for this purpose. Lath is nailed to wood studs and to "nailable" metal studs or may be attached with screws, clips, or staples.

One bundle of six lath units covers 32 sq ft and costs about $3.25 per square yard installed at a rate of about 10 sq yd per hour when nailed in place on walls. For ceiling installation add about $0.50 to $0.60 per square yard. Moisture-resistant lath is often used in bathrooms, kitchens, and other damp locations and is usually supplied in 48″ × 96″ sheets, ½″ or ⅝″ thick. Installation is a bit slower and material costs a bit higher, to give an average cost of about $4.00 to $4.50 per square yard in place. Installation requires about 7½ lb of nails per 100 sq yd or area.

**10.4 PLASTER WORK**  Plaster is a "wet" building material. Job-site mixed of ground gypsum, cement, sand, some lime, pigment, and water, plaster is applied as a semiliquid or a wet paste. Most plaster is factory mixed as a dry powder and may contain small quantities of hair or similar material as a binder. Base coat material is not normally colored, but most finish plaster is pastel colored in the factory mix process. Interior plaster is usually composed of gypsum, while exterior plaster is cement-based compound. Finish surfaces may be almost as smooth as glass using a material known as "Keene's cement," may have various troweled or tooled finishes, or may have a heavy aggregate applied or exposed. Plaster is a multicoat application, usually three coats: a scratch coat used to fill the openings in metal lath and to form a base for the second or "brown" coat, which is a leveling application, and a finish coat, for a total thickness of about ¾″, including the lath. Exterior plaster or "stucco" is a three-coat application over stucco mesh, with a total thickness of about 1″. The finishing coat in most plastering projects is ⅛″ thick, contains the color if any, and is applied to give the surface texture desired.

Plaster may be applied over a number of different base surfaces, including metal lath, gypsum board lath, stucco mesh, concrete or masonry, or solid gypsum plank walls. Most of these occur in residential work as well as commercial and industrial projects, with the possible exception of solid gypsum walls. Thickness of grounds and trim will determine the amounts of plaster required, but ¾″ grounds for metal lath and ⅜″ grounds for gypsum lath are usual. The amount of gypsum plaster varies according to the base material that is used, with about 1000 lb required for gypsum lath per 100 sq yd, about 1200 lb when applied over concrete or masonry, and from 1500 to 2000 lb if used with metal lath. Installation on walls may be done by hand or by spraying and under most union rules requires a helper or "tender" for each plasterer.

Average application of gypsum plaster is 75 to 80 sq yd per day with hand application by a four-man crew and costs about $12 per square yard for three coats in place. Exterior cement stucco costs about $13 per square yard on masonry or concrete and as much as $20 per square yard on wood frame construction. "Thin-coat" plaster used to cover a defaced surface finish can be applied at a rate of about 350 sq yd per day by one installer at a cost of about $3 per square yard. Base coats for tile are usually scratch coat only and also cost about $3 per square yard in addition to the lath base. Moldings, cast plaster wreaths, rosettes, and flowers, which used to be a prominent item in plaster work, are now a nearly forgotten art. Occasionally, an "old-timer" can be found to do this work and costs may run from $3 to $25 or more per lineal foot, depending on the skill required.

**10.5 VENEER PLASTER**  Normal plastered walls using two-coat or three-coat application systems provide fire-resistant walls but remain wet for several days after installation, so require quite a bit of time to dry satisfactorily.

Drywall has some disadvantages also in regard to surface texture, nail popping, and the simple fact that it is not "lath and plaster." To eliminate most of these objections a "veneer plaster" system using a special gypsum panel base plus a thin coat of rapid-drying plaster is now available. Principal advantages are that the base material is installed in large sheets and may almost immediately be covered with a very thin coat of special plaster which may be textured or colored the same as normal plaster. The thin coat of plaster dries very rapidly so may be painted within 24 hr or less after application. Base gypsum board as applied in the same manner and at the same rate as for regular panel work. Veneer plaster is applied approximately $\frac{1}{8}''$ thick by one man at a rate of 350 sq yd per day and at a cost of about $2.50 per square yard for plaster only.

## 10.6 GYPSUM BOARD DRYWALL

"Drywall" originally meant any type of wall material that was applied without moisture, but with further development of a variety of wood, plastic, and other materials has become almost synonymous with gypsum panels. Gypsum panels are fabricated at a manufacturing plant as a sandwich of a solid gypsum core covered on each side with paper or foil. The two principal forms for this material are gypsum lath for use as a plaster base and gypsum panels used as a finishing material. Gypsum lath is not considered drywall but only as a base for wet plaster.

Standard gypsum panels are supplied in $48'' \times 96''$ size, normally $\frac{1}{2}''$ or $\frac{5}{8}''$ thick, and in a variety of tapered, bevel, tongue-and-groove, or square edges. Panels $\frac{1}{4}''$ and $\frac{3}{8}''$ thick are also available but are generally used for double-layer work, overlays, or repair work. The most common surface is plain manila paper, but aluminum foil for use as a moisture barrier and a chemically treated paper face for use in ceramic tile installation are also available. Special types include a gypsum core treated with asphalt and covered with water-repellent paper for use as sheathing, gypsum core panels faced with pattern vinyl, special laminated "fire-code" panels, and high-strength underlayment panels for use over subfloors. Cutting of board is usually done by making a deep score mark and breaking the panel at this point.

Drywall gypsum panels are nailed, screwed, or stapled to wood framing and may be nailed to special studs or screwed into place. Nails or screws are set slightly below the surface of the gypsum board to provide a "dimple." Board edges should be fitted snuggly but not forced into place. An experienced installer should be able to erect about 800 sq ft of $\frac{1}{2}''$ board per day in average-size rooms but may be reduced to one-half that amount in small rooms where cutting and fitting or minimum work space is a problem. Where drywall is used to "fireproof" structural members, installation for a two-man crew varies from as little as 200 sq ft for three layers to a possible maximum of about 500 sq ft for large members. Average cost for $\frac{5}{8}''$ gypsum board nailed to wood studs using a two-man crew is about $0.40 per square foot, not including taped joints.

Joints are covered with perforated paper tape installed in joint compound and should be estimated at about 400 lin ft of tape and 75 lb of prepared joint compound per 1000 sq ft of area. Additional joint compound is used to fill fastener dimples and to smooth the tape surface after embedment. After all joints are filled and dry the surface should be sanded lightly to provide an invisible joint-free uniform plane. Framing must be more accurately installed than for plaster work, as the gypsum board will tend to closely follow an irregular stud wall, thus producing a wavy wall surface. Care in sanding is also important, as too vigorous sanding of joints may produce a shiny smooth spot that is hard to conceal with paint. Taping costs about $0.20 per square foot of gypsum board area installed at a rate of 200 to 250 lin ft per day. Cement may be installed with a putty knife in hard-to-reach areas but in most cases is installed with a pressure gun similar to that used for caulking.

## 10.7 OTHER TYPES OF DRYWALL

Although gypsum drywall is the most common type of installation, there are a number of other materials that do not require water for their installation. Most prominent of these are the various kinds and species of plywood available in many patterns and either finished or unfinished. The actual wood may be the exposed surface, or an applied wood photo covered with plastic may be used. Panels are attached with concealed nails at tongue-and-groove edges or may be carefully face-nailed or glued to the stud supports. If hardwood face material is required, it is normally not required to be "sequence-matched" (see Figure 10–3) but will probably all come from the same veneer flitch. Prefinished plywood, ¼″ thick, with birch-exposed surface will cost about $1.50 to $1.60 averge per square foot in place installed at 400 to 450 sq ft per day. Unfinished birch plywood or knotty pine, A grade, ¼″ thick, costs about $1.75 to $2.00 in place installed at about the same rate as for prefinished material.

With the recent interest in "all-wood" construction, especially for interior walls and ceilings, there has also been a demand for pine and redwood boards and in some locations for cedar or other easily

RANDOM MATCH                    SEQUENCE MATCH

**Figure 10-3**   Veneer matching.

available woods. These are usually supplied in boards from 3″ to about 8″ wide, nominal ¾″ thick, and installed by nailing to studs at a rate of about 300 to 350 sq ft per day for a crew of two carpenters. Boards are most often mismatched deliberately or may be installed from random widths and lengths. Cost in place is about $3.25 per square foot. Where aromatic cedar is used in closets it is usually supplied in bundles, including stock from 1½″ to 2½″ wide, ¼″ to ⅜″ thick, and in random lengths to about 4 ft. Installation is made with nails or adhesive at a rate of 250 sq ft per day at a cost of $2.50 to $3.00 per square foot in place.

Hardboard of various types is also used for wall covering and may be natural faced or covered with a plastic film. Material is usually supplied in sheets 48″ × 96″ × ¼″ thick and is installed at about 500 sq ft per day with a two-man crew. Cost for untempered hardboard, natural faced, is about $1.00 per square foot in place and about $1.50 per square foot for plastic-faced board in place.

## PROBLEMS

**Sample problem:** Estimate the amount of ⅜″-thick gypsum board and the amount of joint cement required to cover the walls and ceiling of a room 12′-0″ × 23′-6″ with an 8′-0″ wall height. The end walls have one 4′-6″ × 4′-2″ window and one 6′-0″ × 4′-2″ window. One door 3′-0″ × 6′-8″ and one door 2′-8″ × 6′-8″ are on opposite long walls.

Walls:   12.0 × 23.5 × 2 = 71 lin ft × 8.0 = 568 sq ft
Wall openings:   4.5  × 4.16   18.72 sq ft
6.0  × 4.16   24.96 sq ft
3.0  × 6.33   18.99 sq ft
2.33 × 6.33   14.75 sq ft
77.42 sq ft

Ceiling:   12.0 × 23.5 = 282 sq ft

Walls:          568 sq ft
  Less openings:   78 sq ft
490 sq ft
Ceiling:       282 sq ft
772 sq ft

Now consider how 4′ × 8′ boards will be used installed horizontally on the walls two 48″ boards wide for an 8′-0″ height. Ceiling to be installed with 8′-0″ length parallel to long walls.

12.0 walls require the full board and two
  ½ boards each wall                                      3 panels

23.5 walls require three full boards on string plus two

full and two ½ boards for second string          6 panels

9 panels

× 2 for additional sides and ends  =  18 boards

(½ boards are required to break the joint locations)

Ceiling: 3 wide × 3 long = 9 boards          9 panels

(8 full boards plus two ½ boards)     Total required     27 panels

48″ × 96″ × 27″ boards          864 sq ft

Lineal feet of joint

| | | |
|---|---|---|
| Each wall top at ceiling | 12.0 × 3 × 2 | 72 lin ft |
| Middle at abutted boards | 24.0 × 3 × 2 | 144 lin ft |
| bottom at floor | | |

Ceiling: 24.0″ × 2 lengths wide plus

12.0 × 2 crosswise          72 lin ft

288 lin ft

Compound required: 1 gallon will spread about 1000 lin ft, so approximately 25 lb will be required for 288 lin ft

Disregard openings, as boards will not cut properly for use of waste pieces.

**10.1** Estimate the amount of wood panel work required for a wainscot 4′-6″ high on all walls of the room in the sample problem. Can you think of a way to minimize waste by a slight change in requirements?

**10.2** Estimate the amount of lathing and plaster required for a bathroom with a wall 5′-0″ wide by 7′-6″ long and 8′-0″ high. Will there be any change if tile is to be installed around the tub?

**10.3** Using the house plans included in this book, beginning on page 194, estimate the quantity of material and the approximate cost for installation of the birch plywood wall in the living room.

# chapter

# 11

# Paint
# and
# Wall Covering

In terms of the construction industry "paint" includes a great variety of liquid finishes applied over wood, metal, concrete, or other materials. Included under this heading are various stains, fillers, varnish, paints, enamels, synthetic liquid plastics, wax, and similar products. Special materials to be used for protection of industrial installations for heat, cold, moisture, acid, or other unusual conditions are usually not included as "paint" but are considered as "special coatings" or "industrial finishes" and are not common in estimating residential painting costs. Application of paint products may be by brushing, rollers, spray equipment, or any combination and is rather strictly controlled by union rules.

Painted protective finishes are usually required on most surfaces, either for exterior or interior exposure. Earlier painting requirements specified that the paint be job mixed using linseed oil, white lead, pigment, and tube colors. Today's paint is almost without exception a factory-mixed product and tinted to the color desired. Very few modern paints use oil-and-lead mixtures as a base material. Most paints are classified by their vehicles (liquids) or binders and include oil-modified resins called *alkyds*, petroleums or *asphalts*, one- or two-part *epoxies*, water-based synthetic *latex*, oil-modified resins called *oleoresins,* heat-curing *silicone alkyds,* evaporation-cured *urethanes,* plasticized copolymer *vinyls,* and a number of others. Quality and recommended use varies widely.

In addition to the actual application of the paint material the project specifications often require work items that may be easily overlooked by the inexperienced estimator. Sanding of wood surfaces to remove machine marks is normally not specified but is supposedly included as "preparation of the surface to the satisfaction of the paint installer." This simply means *sand it* before the first coat and perhaps also between coats. "Protection of adjacent areas" means use drop cloths for floors and perhaps paper taped in place on vertical surfaces. Where finishing nails are "set" or recessed by the carpenter, the painter is required to fill the resulting small depressions with putty "colored to match the adjacent surfaces." Scaffolds, ladders, swing scaffolds, and all other painting equipment is to be automatically supplied by the painter and is seldom specified.

Fire protection during paint application is another item seldom specified. Most times the painter is required, and may do so automatically, to clean brushes, mix paint, and similar work, in the open air. Brushes should be stored in closed containers overnight, *not* in the paint. Fire protection may be nonexistent but may also be a good fire extinguisher readily available, *not* stored in the painter's truck. Proper ventilation for application and drying may mean operation of the heating system or open windows, but the latter may be only temporary during application in order to prevent dust marring the painted surfaces.

## 11.1 PREPARATION OF SURFACES

Sanding and putty work should average about 1 hr per 100 sq ft of flat surface such as exterior trim and about the same amount of time for all surfaces of average wood doors. Sanding and putty work on interior trim or between coats or finish require about 20 to 30 minutes per 100 sq ft, as most of the wood materials used for interior work have already been sanded when manufactured or applied. Since painting is normally estimated in surface square feet it may sometimes be more convenient to consider costs of sanding and putty work at about 50% of the area rather than in 100-sq ft units. If painting is to be done over older, previously painted surfaces it may be necessary to remove the old paint by using a blowtorch and hand scraper at about 4 hr for each 100 sq ft of plain surface.

## 11.2 WOOD SURFACES

Wood surfaces may be classified as *exterior* or *interior* and as *rough* or *surfaced*. In addition, some types of paint products may be installed over damp surfaces but most require a clean dry base. Stain is often used to produce a uniform color or to change the color of the original material and may be oil-base or water-base in a considerable selection of colors. Most stains cover about 250 sq ft per gallon on heavy timber and about 500 to 600 sq ft on interior wood finish. When covered with a coat of sealer, usually shellac, this costs about $0.50 per square foot, including labor. "Solid color" stains have greater hiding power than transparent stains and have a covering capacity of from 200 sq ft per gallon on wood shingles to about 400 sq ft on smooth siding.

Exterior paint may be alkyd resin, oil latex, or full latex-base material. Most exterior paint is classed as "gloss" and is used on new or previously painted surfaces. A wide choice of color is available and coverage of about 400 sq ft per gallon is usual. Primer should be used on all new work. Primer plus two coats of finish paint should cost about $0.45 to $0.50 per square foot applied at 500 sq ft per day by one painter. Interior paint may have a wider range of bases and a wider range of finish materials. In many cases door and window trim and similar moldings are painted with a semigloss enamel while the adjacent wall may have a flat paint finish. In kitchens and baths all surfaces may have an enamel undercoater plus one or two coats of alkyd enamel. Undercoaters or primers cover about 300 to 400 sq ft per gallon. Flat paints and most enamels brush out at about 350 to 400 sq ft per gallon. Primer plus one coat of semigloss enamel is about $0.30 per square foot at 700 sq ft per day for labor of one painter. Enamel undercoater plus one coat of enamel costs about twice that amount. Paint finish on cabinets should be estimated at $0.50 per square foot of painted surface.

Doors and windows present a special problem for painting since there are so many sizes and styles. Many estimators disregard any doors or windows but consider that these areas are solid along with the adjacent wall surfaces. This may be reasonably correct for slab doors and simple windows but will probably be shortsighted for panel doors or windows with mullions and muntins. Other estimators advocate adding a constant size to all openings, say 2'-0", to each width and height. This will allow for painting of edges for doors and muntins for windows. Still another method suggests a constant of perhaps 100 sq ft for all residential doors, 50 sq ft of area to be considered for each side, regardless of door size. A fourth approach is to multiply the area of a window by 2½ times and the area of a door by two times. Using whatever method you desire, a good painter should be able to paint about 125 sq ft of doors or windows per hour.

## 11.3 DRYWALL AND PLASTER

Residential interior walls, as well as most commercial interior walls, are either *wet* walls of plaster or *dry* walls of gypsum board or similar material, as outlined in Chapter 10. In either case these walls usually require an additional finish and paint is used most often. Since plaster is a *wet* material it also tends to retain the moisture required to mix it, and although plaster walls may look dry, they are often damp enough to cause defects in a painted surface. To combat this condition the painter should treat the plaster surface with a solution made from 2 lb of commercial zinc sulfate dissolved in 1 gal of water. Both wet and dry wall surfaces need a prime coat or sealer coat before any additional finish coats are applied.

Commercial primers are available at about $6 per gallon to cover about 400 sq ft, and one painter should be able to prime about 1500 to 1600 sq ft per day. Over this prime coat either one or two coats of finish paint are applied and could be a flat latex, semigloss enamel, or high-gloss enamel, depending on the room location of the wall. Paint on walls, applied with a brush by one painter, requires about 1 hr per 100 sq ft of surface at an average cost of $0.35 to $0.40 per square foot for primer and for two coats of finish. If rollers are used, the painter should be able to cover about 175 to 200 sq ft per hour. There is usually some hand brush work to be done at moldings, corners, and similar locations at about 75 to 100 lin ft, 6″ wide, per hour in addition to the time required for the roller work.

Exterior plaster or stucco is usually not painted when new since most exterior finish coats contain integral color added at the manufacturing plant. When the color fades, the plaster cracks, or the owner simply grows tired of the original color, the exterior plaster may be painted. New plaster should have the same zinc sulfate solution used to help determine dampness, but old plaster has been dried and weathered so does not need this treatment. Most plaster surfaces should receive a prime or sealer coat and at least one full finish coat with average coverage of about 100 sq ft per gallon for "cement-based" paint or 300 sq ft for an oil-base paint. In-place costs for primer plus one coat of finish should be about the same as for interiors, $0.35 to $0.40 per square foot.

## 11.4 METAL SURFACES

Metal used on any construction project may be structural steel-rolled shapes such as beams, angles, channels, and composite shapes or as sheet metal in the form of flashings, gutters, downspouts, louvers, and other light-gauge parts. Most metal must be cut, shaped, and otherwise fabricated and then either painted or given a zinc-galvanized coating. Painted structural steel usually has one shop-applied coat of red lead and oil paint at a cost of about $0.25 per square foot. Tables are available that will give the actual surface area of various shapes and may be used to advantage here. Sheet metal may be supplied as steel sheets, copper, zinc, stainless steel, or possibly lead. Steel-based sheets may be galvanized or black for painting. Steel windows, steel doors, stamped steel louvers, and similar parts are almost always supplied to the job site prime coated. Stainless steel is rarely painted, and copper, zinc, or lead do not need paint protection, due to their composition. Before shop-primed metal parts are finish painted, an indeterminate amount of "touch-up" painting may be required to cover scratches and other blemishes, estimated at about 10% of the area to be painted. Galvanized steel surfaces that have not been prime painted should be cleaned with vinegar or a commercial cleaner to remove the oil remaining before prime coating.

Steel windows are a major item of metal which requires additional finish and will cost about $3 for material for each 100 sq ft of surface and 3 hr of painter time per 100 sq ft. Normally, structural steel shapes are only prime painted when they are concealed but may be painted two or three coats when exposed. Surface area tables should give the square feet to be covered in each shape or may be "estimated" at about 5 to 7 tons of steel painted in 8 hr, or 800 to 1000 sq ft. Steel sheet metal flashings, gutters, and downspouts will require about 1 hr per 500 sq ft of surface while on the ground (not installed, so subject to handling) and only about 250 sq ft per hour when already installed. In estimating the area of sheet metal parts it is customary to multiply the actual area by 2, since most of these shapes are not flat surfaces.

## 11.5 CONCRETE AND MASONRY

Concrete and masonry are usually considered as structural parts and are not painted, especially as they occur in residential work. Compute the surfaces of concrete or masonry as the actual surface area but do not deduct openings of less than 75 sq ft. In most residential projects this simply means to disregard all openings. All joints in masonry should be completely filled with no open spots. This may present a problem when concrete units or brick are laid with "squeezed" joints, giving an overhanging blob of mortar at joints or where the joints are "raked" behind the face of the wall. Either of these joint types is an invitation for water leakage into the building, so waterproofing is of major concern.

New concrete may require a zinc sulfate solution treatment to test it for dampness. New brick work may show signs of efflorescence, usually a white powder-like deposit on the brick face, so may have to be sandblasted or wire-brushed before any finish is applied. First coat on concrete or masonry surfaces should be a "block-filler" type of primer with a covering capacity of about 75 to 125 sq ft per gallon, depending on the porosity of the material. Additional coats may be latex-base or oil-base, with a coverage of about 400 sq ft per gallon. Labor for first coat plus one coat of finish should require about 1 hr per 100 sq ft for the prime coat but only about 30 to 40 minutes per 100 sq ft for the finish coat. Concrete floors may be chemically stained or surface painted. If stain is used, the finish will be varigated due to chemical reaction, with the inert cement remaining in the slab. A sealer coat is necessary over the stained surface and the stain and sealer costs about $1.60 per square foot. Floor enamel is a "paint" specially prepared for this use, with a coverage of 400 to 500 sq ft per gallon applied with labor at about 20 to 30 minutes for 100 sq ft.

## 11.6 MISCELLANEOUS PAINTING

Wood floors and some other wood surfaces are normally sanded smooth, then stained, and a filler coat applied to fill the open pores of the wood. This is followed with clear or orange shellac which is rapid-drying sealer, and further finished with varnish or urethane. Although this finishing process is often done as a part of the wood floor installation, it may also be a part of the painting operation and should cost in the area of $0.90 to $1.00 per square foot complete.

When varnish or urethane is used on wood which requires a clear finish, the painter should be able to cover about 800 to 1000 sq ft per day with one coat. Varnish gives about 550 sq ft of coverage per gallon and is available in either gloss or semigloss sheen. Sealer plus one coat of varnish on wood should cost about $0.25 per square foot. Where extreme exposure to water or sun are expected, a marine "spar" varnish should be used at slightly higher cost.

Heated surfaces such as radiators, grills, boilers, stacks, and similar areas require a heat-resistant paint specially prepared for this use. Aluminum flake pigment paint with a spread rate of about 600 sq ft per gallon may be used as a sealer or prime coat or as a finish material. This material is heat resistant from about 400 °F to as much as 1000 °F, depending on the formulation. "Stack paint" in a few neutral colors is also available, with a coverage of 500 sq ft per gallon and is heat resistant to about 500 °F.

## 11.7 WALL COVERING

Wall covering is primarily of two major types: wallpaper or vinyl composition sheet material. Wallpaper is available in single rolls or double rolls, 18″ or 20″ wide, and is figured and priced as single rolls containing 36 sq ft or 4 sq yd. Calculate the required area as the full area of the walls less any openings and add 20% for waste and average pattern matching. DO NOT FORGET THE CEILINGS! Wallpaper paste, wheat flour mixed with water, and glue size to seal the walls are two major items that may be overlooked. Wallpaper paste is available in powder form to be job mixed with water and 1 lb of powder will make about 2 gal of paste at a cost of about $1.25. One gallon of prepared paste will be required to hang 12 single rolls of medium-weight paper. Glue size consists of flake glue dissolved in water and will cover about 650 sq ft per gallon and requires about 30 minutes of labor for each 100 sq ft. The cost of wallpaper varies from as little as $0.50 to as much as $5.00 per single roll, depending on weight, pattern, repeat sequence of pattern, and surface texture. A wallpaper hanger should be able to hang approximately 30 single rolls in 8 hr.

Wall covering other than wallpaper is of two major types: very thin wood veneer glued to a backing, or vinyl-surfaced material either with or without a backing. Wood-faced material is available in about 75 different woods and patterns, with material ranging from $0.75 to $5.00 or more per square foot, is manufactured in widths to 24″ and lengths to 12 ft, and is usually prefinished. Installation requires expert workmanship and can be estimated at about 100 sq ft per day. If lacquer sealer is applied, figure coverage at 350 sq ft per gallon at about 100 sq ft per hour. Vinyl wall covering is composed of a woven fabric combined with a vinyl resin face material and is available in widths of 54″ to 30 yd in length. There are literally thousands of patterns available in three general weights of material, from 17.5 oz per lineal yard to 30 oz per lineal yard. Estimating quantities of covering, glue size, and adhesive are accomplished in the same manner as for wallpaper and a competent paper hanger should average about 400 sq ft per day.

## PROBLEMS

**Sample problem:** Estimate the quantities required to install the following paint and wallpaper. The living room is 15′-0″ × 25′-6″, 8′-0″ ceiling. One steel window on the short side is 6′-0″ × 4′-2″, one door on the long side is 2'-8" × 6'-8", and one sliding aluminum-framed glass door is 8'-0" × 7'-2" on the other short side. Required to paint sealer on wallboard plus two coats of latex on all surfaces including ceiling, except that one 25'-6" wall with a 2'-8" door is to be wallpapered.

Walls: 15.0, 15.0, 25.5 =       55.5 lin ft    Ceiling 25.5 × 15.0 = 382.5 sq ft
                                      ×   8.0 height
                                      444.0 sq ft

Disregard door and window openings, as they require additional painting as below:

Window:  6.0 × 4.16 × 2 = 50.0 sq ft      Walls:       444.0 sq ft
Door:      8.6 × 7.16 aluminum not painted   Ceiling:    382.5 sq ft
                                             Window:      50.0 sq ft
                                                          876.5 sq ft

Wallpaper: 25.5 × 8.0     204.0 sq ft    Paint req'd:
Door:        2.66 × 6.66    17.7 sq ft    Sealer at 400 sq ft    ¼ gal
      Net area            196.3 sq ft    Finish at 400 sq ft
                                             2 coats          4½ gal

Assume 18" wide paper     38.25 strips
                        ×   8.0
                        306.0 lin ft

$$\frac{\text{Say 200 sq ft}}{36 \ \text{sq ft single roll}} = 5.5 \text{ single rolls}$$   Order 6 single rolls or 3 double rolls

Glue size required:  ½ gal
Wallpaper paste:  ½ gal

**11–1.** Estimate the required amount of sealer and enamel for a bathroom 5'-2" × 8'-0" with 7'-11" ceiling height, with a 60" × 30" × 18" high tub across one short end, with one wood casement window above the tub 4'-0" × 1'-6" and door opposite 2'-0" × 6'-8".

**11–2.** Determine the quantity of vinyl wall covering and adhesive and the approximate cost installed for an entry 4'-6" × 6'-0", with 3'0" × 7'-0" entry door, closet door 2'-8" × 6'-8" on one 6 ft wall. The wall opposite the entry door is "cased" without a door.

**11–3.** Estimate the quantity of paint required for all exterior surfasces for the house plans included in this book. Exterior paint will be one coat of sealer plus one finish coat. Metal parts are all galvanized without primer coat.

# chapter

# 12

# Tile
# Work

Ceramic tile is one of the oldest building materials known and in one form or another was used as early as 500 B.C. Tile is a kiln-fired clay product that may be glazed or nonglazed mosiac, in dozens of different shapes and sizes, and installed in a number of different ways. In ancient times ceramic tile was used for floors, walls, ceilings, as decoration, and in almost any way that a person could think of. Today ceramic tile is still used on floors, walls, and some ceilings as well as countertops and in a number of other special locations. Some tile is used for exterior surfaces of commercial work and a great number of special types are available for commercial kitchens, bakeries, hospital laboratory rooms, and surgical spaces, and for areas subject to acid or corrosive liquids. Ceramic tile is wear and scrub resistant, colored or available in hand-painted patterns, reasonably permanent, and not over costly.

## 12.1 GLAZED WALL TILE

Glazed ceramic tile is available in standard sizes of 4¼" × 4¼", 3" × 6", 4¼" × 6", and a number of other sizes to 6" × 6", nominally ⅜" thick, and may have a high-gloss glaze or a satin or "matte" glaze in color. Glazed tile is generally used for walls or ceilings but not for floor installation, as the glazed surface tends to become very slippery when wet. Where glazed tile is used for floors the glaze may be a matte glaze or may contain abrasive particles to reduce the danger of slipping. The edges of standard glazed tiles may be square cut or "cushion edge" (edges slightly beveled) and may or may not have spacing lugs to ensure uniform joints. Spacing lugs are small projections along the edges that help to keep adjacent tiles separated by the correct amount and are covered with the joint grout when fully installed.

Glazed tile is installed by one of three standard methods: by full mortar beds, by "dry-set" thin cement grout, or by organic adhesives (Figure 12–1). Most wall tile is supplied in boxes of single units and are installed piece by piece. For installation in a full mortar bed the pieces are first thoroughly wet, then installed in a wet mortar over a plaster or other solid backing. For wood stud or metal stud walls a waterproof barrier (15-lb asphalt-saturated paper) is applied with metal lath and cement plaster scratch coat leveled off to allow about ¾" of mortar over the scratch coat. A mortar bed of one part Portland cement, ½ part lime, and 5 parts sand is applied and allowed to set but not dry out. A setting bed or bond coat of neat cement is applied over the still-plastic mortar and the wet tiles installed in the bond coat. Joints are kept open with lugs on the tiles or with a length of heavy fishing twine of the desired diameter. After the tile has set the twine is removed, the joints cleaned, and the grout installed flush with the tile surfaces. Excess grout is removed by cleaning the wall with a wet sponge or cloth.

If the "dry-set" method is to be used, the backing may be masonry, concrete, gypsum board, mortar setting bed, or some forms of tile or marble that are smooth and in a flat plane. Tiles are installed dry in a bed of "dry-set mortar" approximately ⅛" thick.

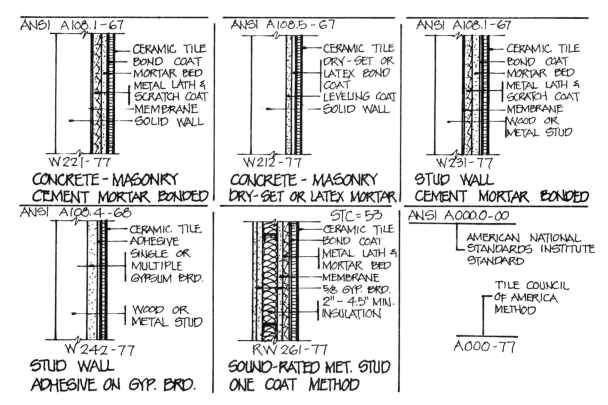

**Figure 12-1**   Ceramic tile wall installation. (Courtesy of the Tile Council of America.)

This mortar bed is not a setting bed and is not intended to be used in truing or leveling the work of others. Grout used for dry-set installation is specially compounded for nonwet tile, but damp curing of the completed walls will probably increase the grout strength.

Adhesive-installed ceramic tile is entirely a dry operation and is usually installed over gypsum board surfaces. Gypsum board should be of the water-resistant type and may also be fire-resistant or sound-absorbing type installed over wood or metal studs. Multiple layers of gypsum board may be necessary to provide a minimum of ½″ thickness. Organic adhesives are factory compounded and are applied with a notched trowel in one coat into which the tiles are set. Tiles are set dry. Allow a minimum of 24 hr after the tiles are set for solvent to evaporate before grouting joints. Grout the same as for other types of installation. This type of installation is not recommended for exterior tile work.

## 12.2 MOSAIC FLOOR TILE

Mosaic tile is designed primarily for installation as a floor surfacing but may also be used for walls if desired. A great number of patterns are available, with individual units ½″ × ½″, ½″ × 1″, 1″ round, 2″ round, hexagonal, and various combinations, unglazed, natural color clay ranging from red through browns and blue-greens, and a variety of blends and mixtures. Most mosaic tile is supplied in 12″ × 12″ patterns glued to a paper applied to the face side of the tiles. After installation the paper is soaked off and removed before the tile is grouted. This method eliminates the time-consuming task of installing one tile at a time and presents a more uniform finished surface.

Floor installation may be made upon two typical subfloors, concrete or wood (Figure 12–2). The same three general methods of installation may be used for floor tile as for wall tile, with a few exceptions and/or additions. When installation is over concrete with a full mortar bed, there should be a cleavage membrane of 15-lb asphalt-saturated paper or 4-mil polyethylene film initially placed over the concrete. The mortar bed, ¾″ to 1¼″ thick, is reinforced with metal lath or stucco mesh. The mosaic tile is then placed, paper side uppermost, in the setting bed and pounded into the required plane. Special care should be exercised where a pitch in the floor is required and expansion joints should be installed over any structural members, and at areas of about 6 sq ft. After the tile has set, the paper backing is removed and the tile is grouted.

When mosaic floor tile is installed over a wood subfloor, the usage and traffic expected is a big factor. Where floor tile is going to be installed in a full mortar bed, all of the same requirements for installation over concrete are necessary: cleavage membrane, reinforcing, mortar bed, and bond coat. In addition, the subfloor may have to be dropped between the joist to allow for the full depth of the mortar bed. Installation with expoxy mortar and/or adhesive requires a double layer of wood flooring and should be limited to residential use only. Two layers of ⅝″ to 1″ thick plywood are

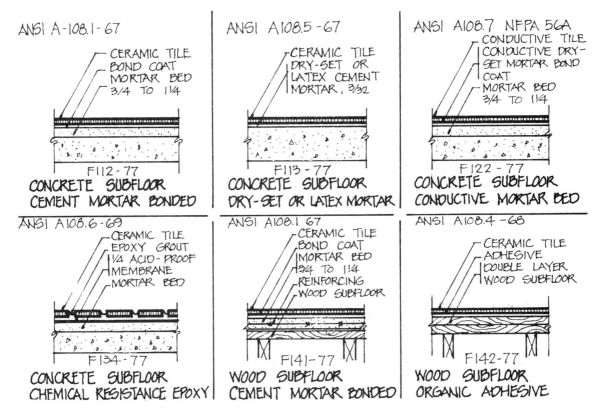

**Figure 12-2** Ceramic tile floor installation. (Courtesy of the Tile Council of America.)

recommended and this is usually *not* indicated on the drawings, so the estimator must keep it in mind when taking off the subfloor material beneath the tile. The remaining portion of thin-set or adhesive-installed tile is the same as for a concrete subfloor.

## 12.3 QUARRY TILE

Until fairly recently, quarry tile has been used primarily in commercial installations but is now appearing more in residential construction. Quarry tile is normally available in unglazed natural red colors in sizes from 4″ × 6″, 6″ × 6″, to 12″ × 12″, ½″ thick. Several firms also manufacture small quarry tiles, 1″ × 1″ × ¼″ or 2″ × 2″ × ¼″, in plain or blended color patterns, mounted on 12″ × 12″ sheets similar to mosaic tile. Installation of quarry tile is seldom made on wall surfaces except perhaps in special conditions for commercial kitchens, packing houses, and similar locations. Bases are available and may be a part of the quarry tile floor installation work. Installation of quarry tile is made in the same manner as for mosaic tile over either concrete or wood subfloors. Where chemical or acid resistance is required, the grout may be furan resin, which requires special attention and additional cost but is not usually necessary for residential construction.

## 12.4 CERAMIC TRIM AND ACCESSORIES

In addition to the standard flat shapes and sizes of ceramic tile there are complete lines of trim pieces in matching colors and material (Figure 12–3). These include straight and coved bases, coved and

127

bullnose corners, square and bullnose cap members, counter edges, and a number of related pieces. There are also shapes for swimming pools and tiled-in baths and showers. Ceramic bath accessories such as paper holders, towel bar brackets, and soap dishes, either recessed or surface mounted, are also available and may often be included as part of ceramic tile work.

Decorative tiles are referred to as "hand-painted," "Italian," "Mexican," or by a number of other names that indicate a surface-applied design in various colors or glazes. These tiles are available in 6″ × 6″ × ¼″, 12″ × 12″ × ½″, and several odd sizes, in hundreds of patterns and colors. As they are really hand-painted, no two will be exactly alike, but the general pattern and colors will blend into a fairly uniform appearance. Some trim members are also available, as are sculptured and raised surfaces. These decorative tiles are

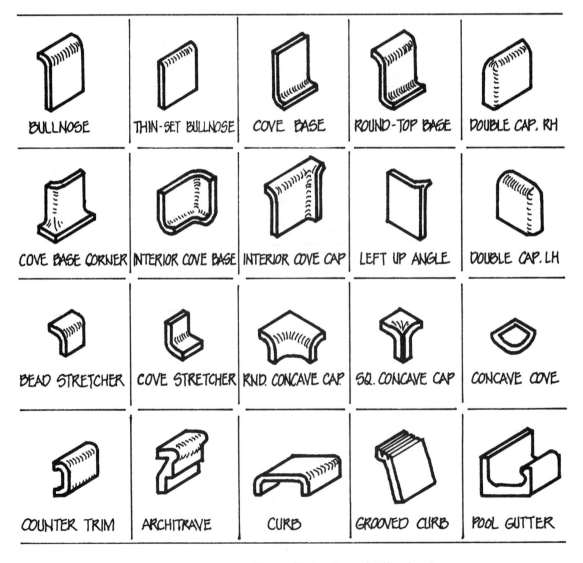

**Figure 12-3** Ceramic tile trim shapes.

expensive and priced according to design and color and in most cases are used only for accent.

## 12.5 COLORED CEMENT TILE

Cement tiles are also becoming more available and more widely used, especially in the southwestern United States, where the source, Mexico, is nearer at hand. Cement tiles are usually about 12″ × 12″ × ½″ or 1″ thick, some standard patterns but mostly any pattern that the maker had at hand. The base of the tile is a mixture of cement, sand, water, and perhaps a little fine gravel, topped with a layer of cement grout about ⅛″ to ¼″ thick. The grout is colored and any pattern is also colored. These tiles are made with a metal form so the pattern generally is fairly uniform, but the colors are added by hand so may vary depending on the tile maker. If special designs are desired, the form may be changed and almost any pattern is possible. Most installations of cement tile are upon concrete sub-floors or in patios and similar locations, as the tiles weigh considerably more than ceramic tiles. Installation is in cement mortar with cement-grouted joints.

## 12.6 INSTALLATION COSTS

Almost all of the tile installation is done with a two-man "crew," one tile setter and one tender. The estimator should be aware that where full mortar setting or *dry-set* installation is required, the reinforced plaster or mortar bed may be a part of the tile or may be included as plastering work and should be estimated accordingly. Cement plaster scratch coat for tile installation should be installed at a rate of about 300 sq yd per day at a cost of about $2.75 per square yard. Tile installation is figured by the square foot of wall, floor, or ceiling and by the lineal foot for trim and base. Size of spaces to be tiled and the number of spaces also has a great influence on in-place costs. Material cost will remain constant with the areas concerned, but labor costs will increase as the spaces become smaller or more difficult to install the tile. For installation of wall or floor tile, base, and cap units in areas smaller than 35 sq ft, multiply labor time by 2 and for countertops and sink splashes, multiply the area by 3 for comparable labor costs.

Installation of 4¼″ × 4¼″ glazed wall tile, *set in mortar,* should be at a rate of about 60 sq ft per day and cost about $3.50 to $4.00 per square foot in place. If the same-size tile is installed with the *adhesive-set* method, the in-place cost is about $2.75 to $3.00 per square foot at about 175 sq ft per day. Mosaic floor tile installed in mortar depend somewhat on the type and tile size but should be installed at about 175 to 180 sq ft per day and cost an average of $3.00 to $3.50 per square foot in place. The same tile set with adhesive or thin-set costs $2.75 to $3.00 per square foot. Quarry tile, 6″ × 6″ × ½″, set with mortar at a rate of 100 sq ft per day, costs about $3.25 per square foot in place. Cement tiles 12″ × 12″ × 1″ thick, cost about $1.50 to $3.00 each, depending on design and color, and should be installed at about 50 to 60 units per day at a total in-place cost of about $6 per square foot, including the setting bed.

Trim, base, and cap units set in mortar at a rate of about 75 lin ft per day cost $4.00 to $4.50 per lineal foot, but where the *thin-set* method is used should be about 50% more production at about 25% less cost. Base and trim for quarry tile set in mortar at a rate of 100 lin ft per day should average $3.00 per lineal foot in place. Counter-tops of tile with ceramic edges tile and a 6″ back splash require about 12 hr of crew labor per 100 sq feet and cost approximately $6.25 per square foot in place. Wainscot 6 ft above the rim of a recessed 60″ tub, back, and two ends, set in adhesive, costs about $175 complete in place and requires about 5 hr. Accessories of ceramic or metal average about $5.00 each in cost installed at about 20 minutes each at an in-place cost of $8.00.

## 12.7 PLASTIC AND METAL TILES

Plastic tiles are made of various components in nominal sizes of 4¼″ × 4¼″ and 9″ × 9″ in many colors and patterns. Application may be made over almost any smooth dry surface except insulating board. The base surface should be properly plumb and level, as there is no possibility to correct uneven surfaces with the tile material. Tiles are installed at about 75 to 100 sq feet per day at a cost of about $2.00 to $2.25 per square foot. About 2½ gal of adhesive is required per 100 sq feet and costs about $6 to $8 per gallon.

Metal tiles are formed from aluminum, steel, or copper with a fired-on colored enamel or porcelain finish. Some tiles are available in natural-finish copper or stainless steel, with or without a protective coating. Installation is at about the same rate as for plastic tiles. Metal tiles cost approximately $4 per square foot for 4¼″ × 4¼″ size in enamel color or about $6 per square foot for stainless steel. Trim of various shapes averages about $2 per lineal foot in place.

## 12.8 PRECAST TERRAZZO

Terrazzo is a man-made product of cement, sand, marble chips, coloring pigment, and water placed as a plastic mix. Usually, this compound is placed as a floor material, and when set and dry, is ground smooth and polished. Color is integral due to the pigment and the marble chips and is available in accordance with sample mixes formulated by the National Marble and Mosaic Association. In most normal installations expansion joints of metal are incorporated in the floor at about 48″× 48″ intervals and bases and similar shapes are hand finished.

In modern residential construction there is not much opportunity to install terrazzo by on-site methods. Precast terrazzo is available in tiles 12″ × 12″ × 1″ thick or 16″ × 16″ × 1½″ thick and may be installed over concrete or wood subfloors with a membrane, reinforcement, and mortar bed in a method similar to that required for floor tiles. Stair treads are another precast terrazzo item that may be used in residential work. Precast terrazzo tiles 12″ ×

12″ × 1″ cost about $10 to $12 per square foot in place with trim at about $5 to $6 per lineal foot. Precast stair treads cost about the same per lineal foot as floor tiles cost per square foot.

**12.9 STONE TILES**   Two types of stone may be considered as available "tiles" that could be used for possible residential installation. Marble in its many colors is usually cut and polished for a specific project. However, several companies produce precut, polished, floor tiles of marble in unit sizes about 4″ × 12″ × ⅜″ thick with flat edges for butt joints. Marble is installed in a mortar bed at a rate of about 50 to 60 sq ft per day and costs approximately $8 to $10 in place, depending on the type of marble used.

Natural slate is another stone that may be cut and polished for use as a floor tile. Grade A slate in blue-black natural color is available in 6″ × 6″ to 18″ × 24″ and ½″, ¾″, and 1″ thick. Weight is from about 5 lb per square foot for ¼″ and ⅜″ thickness to about 15 lb per square foot for 1″-thick material. Irregular sizes for random patterns are also available up to approximately 4.0 sq ft per piece. In addition to floor tiles, slate is also used as stair treads, window sills and copings, fireplace hearths, and facings. Slate tiles, 6″ × 6″ × ½″, thin-set at a rate of about 150 sq ft per day cost approximately $3 per square foot in place. Slate laid in random patterns at about the same rate per day will cost about $5 per square foot in place.

## PROBLEMS

**Sample problem:** Using 4¼″ × 4¼″ ceramic tile, estimate the quantity of tile needed to cover the back of a 60″ tub and each end of 30″. Wainscot to be 60″ high. Include lineal feet of bullnose trimmed all edges.

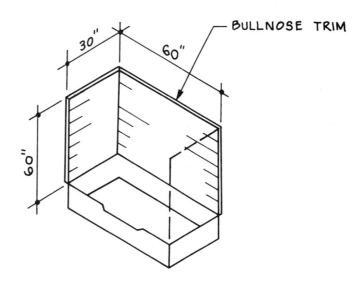

Back tile:    $\dfrac{60'' \times 60''}{4.25 \times 4.25} = \dfrac{3600}{18.06}$    199.33 sq in.

| | | |
|---|---|---|
| 14 tiles vert   = 59½″ | Ends 14 vert = 59½″ | |
| 14 tiles horiz  = 59½″ | 7 horiz = 29¾″ | |
| 14 × 14 tiles = 196 tiles | 98 × 2  ends = 196 tiles | |
| plus 2 ends      + 196 | | |
| Requires        392 tiles | | |

Trim   60″ + 30″ + 30″ + 120″ (height at ends) = 240 lin in.

$$\frac{240 \text{ lin in.}}{12} = 20 \text{ lin ft of trim}$$

**12-1.** Estimate the amount of mosaic tile supplied in 12″ × 12″ sheets, set in a cement mortar bed 1¼″ thick, in a bath 8′-6″ × 10′-0″ using a 60″ × 27″ tube and a 36″ × 30″ adjacent shower unit on the same wall.

**12-2.** Estimate the amount of tile and trim for a countertop 24″ wide, 96″ long, with an "ell" at one end 24″ wide and 36″ long. Consider 6″ of back splash and required lineal feet of edge trim.

**12-3.** Using the house plans included in this book, estimate the cost to install mosaic tile in the bathroom floor area.

# chapter
# 13

# Miscellaneous Items

Despite the seemingly complete separation of the various building materials discussed earlier in this book, there are a number of items that do not easily fall into one subtrade or another. A great many of these items are "buy-outs" by the general contractor, who simply purchases the items and has his own personnel install them. Some of the items are supplied and installed by the manufacturers, while some, discussed in later pages, may be purchased by the owner and installed by the contractor or the owner. A few things are "estimated" as an allowance and are usually installed by carpenter labor.

**13.1 BATH ACCESSORIES**

There is often some confusion regarding what is considered "bath and toilet accessories" and what is considered "plumbing fixtures." In this section only bath and toilet accessories will be discussed. Plumbing fixtures, tubs, water closets, lavatories, sinks, and similar items are *plumbing, not* accessories. Accessories include toilet paper holders, towel bars, soap holders of various types, tumbler and toothbrush holders, robe hooks, and similar items. Also included in this category, and intended primarily for schools, commercial, or industrial installations, are a great many shapes and sizes of equipment designed to provide or dispose of paper towels, sanitary napkins, toilet seat covers, liquid soap, and janitorial supplies. Grab bars of several hundred types and designs are also availabe. Most of the items, except towel bars and grab bars, are manufactured for surface or recessed installation and are provided in stainless steel, chrome-plated finish, white enamel, or special finish on order.

The average residential bathroom requires one toilet paper holder, two soap dishes (one with integral grab bar for tub location), two towel bars 18″ to 24″, and one each of tumbler/toothbrush holder and robe hook. Half-baths require the same except less one towel bar and tub soap dish. Most units will be surface mounted except in upper-level-type construction. Bath accessories are commonly included in a contractor's bid as an "allowance" established by the architect or owner and the exact type is selected after a contract is signed, in a manner similar to that explained for finishing hardware, (Section 8.7). Average cost for most residential bath accessories is $5 and requires about ¼ hr each of carpenter time to install.

## 13.2 BLINDS, SHADES, AND DRAPES

In most construction these items are *not* included as a part of construction costs but are purchased and installed under separate contracts by the owner. However, they are a construction cost item in some projects so will be briefly considered here. First let us make sure that we know the terminology. Blinds are usually of a metal material, aluminum or steel, with applied finishes of enamel or polyester, designed to close horizontally or vertically. Horizontal blinds are most common and consist of shaped slats approximately 2″ wide supported by tapes from a headrail and operated for opening, closing, or tilting by a cord arrangement. These "venetian blinds" are available in a wide variety of patterns and cost from $0.75 per square foot to $5 and can be installed at about 450 sq ft per day, at an average in-place cost of $2.75 for labor.

Shades may be textured cloth, one-piece, operating from a spring roller at the window head, available in numerous patterns and colors, at a cost of $1.00 to $1.50 per square foot in place. Window shades are also available with wood slats or dowels interwoven with colored yarns for pattern. This type of shade usually rolls up into a valance unit by using a cord system, but the openings between slats are not adjustable due to the weaving. In-place costs for this type are about $3 per square foot in place.

Most residential drapes are of cloth, lined with another material or not, and may be treated for fire resistance. These operate from a traverse track at the window head with a system of cords and pulleys. Where blinds and shades are figured for the actual area, drapes are usually pleated along the head and should be estimated at three to five times the width required. In-place costs for lined cloth drapes average about $3 per square foot, with traverse track adding another $5 per lineal foot of track.

## 13.3 KITCHEN APPLIANCES

Every modern kitchen has some, if not all, of the normal appliances built in and considered a part of the construction cost. Kitchen appliances will probably include as a major item a countertop electric range unit and built-in oven, a dishwasher, a laundry washer/dryer, an electric refrigerator of about 12- to 14-cu ft capacity with defrosting included, and a microwave unit at an average in-place cost of about $500 each. In addition, minor items, such as range hoods, disposal units, and ventilation fans, cost about $175 each and a 40-

gal electric water heater, trash compactor, water softener, and similar items cost about $300 per unit in place. Cost of many appliances to speculative builders is based on "package deals" to include ranges, dishwasher, refrigerator, and laundry equipment, which may mean "bottom-of-the-line" quality but which still retains the name of a nationally known manufacturer for sales purposes. Sinks and laundry trays are not considered "appliances" but are plumbing items.

## 13.4 PREFABRICATED FIREPLACES

More and more families now want fireplaces in their homes, both from an energy-saving position and for social enjoyment. Masonry fireplaces constructed at the job site are expensive and if not properly built often smoke or are otherwise undesirable. This situation has created a market for prefabricated free-standing fireplace units of sheet metal in a considerable number of square, oblong, and round designs. The average cost of a unit fireplace is about $350 to $375 and installation will add another $200. Chimney sections of double-pipe form cost about $35 per lineal foot in place. Additional costs for this type of fireplace may be those required for floor or wall coverings required by code for these types of fireplaces. Most codes require a fire-resistant floor beneath the unit and for a distance of at least 12″ on all sides. This flooring may be sheet metal, tile, concrete, or masonry but should be estimated separately. Walls that are nearer than 8″ should also be covered with a fire-retardant material, such as plaster, masonry, or metal, and again need to be considered as possible additional costs.

## 13.5 POSTAL REQUIREMENTS

Every residence, commercial, or industrial location in the United States is required to provide an "easily accessible" depository for postal distribution. In rural areas a metal box of standard size and shape may be adequate, at a cost of about $25. In single residences the most common type is a box installed in the wall with a mail slot outside and a door within, installed at a total cost of about $35. Wall mail slots or those installed in doors, without an interior box, cost about $12 to $15 in place. Apartments are usually provided with multiple-box arrangements installed in the wall, front loaded, key operated, and will average about $45 to $50 per box, depending on the number and the arrangement in each section. Special framing may be required around such boxes, so *double check*!

## 13.6 MEDICINE CABINETS

Almost every bathroom or toilet room requires a medicine cabinet of some sort. Unfortunately, these are often indicated on the drawings by a box-like symbol with "M.C." alongside and no specification regarding size or type. There are many sizes and types, from the most simple surface mounted with mirror door, 16″ × 20″, to the recessed 60″ × 30″ three-compartment hinged or sliding mirror doors and with or without lighting. Many arrangements of mirrors as well as lighting combinations are possible. Finish of medicine cabinets may be white or colored enamel in the more economical units, stainless steel or chrome-plated metal for others. The small 16″ × 20″ recessed, unlighted, white enamel units cost about $50 plus about $10 installation costs. The larger units, with center mirror and two cabinet ends, overhead lighting, stainless steel, cost about three times that of smaller units, or about $150 to $175, and installation is another $60, for a possible total cost of $200 to $250.

## 13.7 SHOWER DOORS

Most residential bathrooms are equipped with a shower unit, either as an addition to the bath tub or as a completely separate fixture. In the latter case the shower may be a prefabricated plastic enclosure complete with door and is considered a plumbing item. Tubs and tiled-in showers, however, require a curtain or other barrier to contain the shower splash, and this is usually not a plumbing item. Shower doors may be sliding type used at tubs, or hinged type used at shower enclosures. Most frames are of anodized aluminum or stainless steel and should have panels of tempered glass, wired glass, or heavy plastic.

Tub enclosures are usually two-door sliding type approximately 60″ long and 4′-6″ to 5′-0″ high. Average cost in place for an aluminum frame with tempered glass is about $250 and requires 4 hr to install and caulk edges. Shower doors are normally about 24″ wide, metal framed with plastic panels. Matching side panels may be required, depending on the size of the shower opening. Hinged doors cost from $125 to $300 each in place depending on the quality, and require about 1½ to 2 hr each to install.

## 13.8 SAUNAS AND HOT TUBS

Saunas and hot tubs in various styles have been used by human cultures for years but have become a feature to be seriously considered in residential construction for only about the last 10 years. Saunas and hot tubs are usually available in kit form either for installation by the manufacturer or by the owner or builder. Saunas are available in kits from about 24 sq ft of floor with a capacity of two persons to a 75-sq ft floor area and a capacity of about 10 persons. Residential sizes are usually about 4′-0″ × 6′-0″ to 5′-6″ × 8′-0″ and 7′-0″ in height, with walls from 2½″ to 4″ thick, fabricated of kiln-dried lumber enclosing a fiberglass or polyurethane insulation core. Heat is normally supplied by an electrical unit from 4 to about 10 kw.

Hot tubs are an even more recent addition than saunas to the residential construction picture and are available in plastic or wood, with redwood a favorite material. Sizes vary from about 48″ in diameter to 96″, with a depth of about 5′-0″. Hot tubs must be set either with the top level with the floor or with a raised platform surround. This of course means that considerable carpentry work may be necessary in addition to that required for the actual tub installation. Heater, pumps, and related equipment require a space of about 3′-0″ × 3′-0″ with concrete slab floor. The tub is installed on a cribbage of beams or concrete piers.

Sauna units, prefabricated including heater and complete installation, approximately 25 sq ft in size, cost about $2000 to $2500 in place and require a full day for a crew of four. Units to 75 sq ft cost about $4000 and about the same time for installation and crew. Hot tubs range from $1000 to $2400 in place. In addition to the cost of these units and any wood platforms or surrounds, do not forget that electrical heaters require considerable power and will therefore require a much larger electrical capacity and its attendant costs than for a residence without these units.

**13.9 SMALL ITEMS**    There are quite a few small or unusual items that may be included in residential work in addition to those noted. The following short list may indicate the wide range of possibilities.

- Attic disappearing stairs
- Door chimes or bells
- Garage door openers
- Fireplace screens
- House numbers

- Inset entrance mats
- Mirror closet doors
- Observation "peek holes" for doors
- Smoke alarms

The cost of these items is minimal and varies with quality and type, but a fair "rule of thumb" regarding in-place costs is to multiply the material cost by 2. At least this estimate will cover most costs.

## PROBLEMS

**13-1.** Estimate the cost for providing cloth shades for full coverage of all windows in the house plans on pages 194-197.

**13-2.** Estimate the cost of all bath and toilet accessories and a middle-cost medicine cabinet for the same house.

**13-3.** Investigate from manufacturers' catalogs or local distributors the approximate cost of at least three of the items mentioned in Section 13.9.

# chapter

# 14

# Plumbing

Plumbing work, as well as heating–ventilating–air conditioning (HVAC) and electrical work, are major subcontracts in almost every construction project. Plumbing work for residences is generally limited to rough plumbing and finish plumbing. Rough plumbing is considered as all of the piping required to supply hot or cold water to the fixtures and to remove the waste from these fixtures. The fixtures and the trim (faucets, fittings) are considered finish plumbing. In general, everything that is concealed is termed "rough" plumbing, with all of the exposed work "finish" plumbing. With these two separate installations, the plumbing is almost a continuous process, from installing the buried pipes to the streets, to installing the piping in the building walls and floors, and finally to installing the fixtures and trim.

Plumbing work is often shown simply as a diagram or in even less completeness, as fixture symbols and nothing more. This means that an estimator must be able to visualize the required work in detail, including all of the fittings and other factors of construction that may cause problems. Types and sizes of piping, code requirements, and related items are almost entirely the problem of the estimator or some other person who may lay out the actual location of the piping. This plumbing is estimated exactly as any other work, including a quantity survey of materials plus labor costs, overhead, and profit. A thorough knowledge of plumbing is required, so much of the nonplumber estimating is done on a cost-per-unit-in-place basis, which may give an idea of costs adequate for some uses.

## 14.1 ROUGH PLUMBING

Rough plumbing is considered as all of the piping needed to provide water supply and waste removal. This rough plumbing includes hot and cold water, the sewer piping, vent stacks, and all the concealed work. It may also include septic tanks if required and connections to municipal water and sewer. Gas piping from the street location to all required outlets is also a part of the rough plumbing work. This piping may be of a number of different materials (cast iron, galvanized steel, copper, plastic, lead, or clay tiles) and of a number of different weights. Fittings include all of the various elbows, tees, unions, nipples, caps, and similar parts to connect the various pipes together. Control valves, plugs, traps, and other fittings are also included.

Sewer lines or "soil" piping is the section of pipe required to carry the waste from the building to the municipal sewer in the street, or in rural areas, from the building to the septic tank. Inside the building this piping is often termed "house drain." The piping used outside the building is cast iron, clay tile, or in some locations may be of plastic. Within the building the same material may be used, with the possible addition of galvanized steel pipe. Vent piping for the sewer system is usually cast iron in older buildings, but in newer construction may be of galvanized steel or plastic. Cast iron pipe is usually hub-and-spigot type, with each joint packed with treated oakum and sealed with molten lead. This type of pipe is available in most common sizes from 2″ to 10″ with 4″ and 6″ most generally used, and in standard lengths of 5′-0″ and 10′-0″. Weight is termed "standard" weight or "extra-heavy."

Water piping may be of galvanized steel, copper, wrought iron, or plastic. Galvanized steel, black iron, and wrought iron are threaded for screwed fittings, while copper usually has sweated joints and plastic is assembled with liquid "glue." Gas lines should be installed only with black iron piping, as most gas chemically reacts with the zinc used for galvanizing and may cause defects in the joints and other parts of the system. Similarly, "hard" water containing an excess amount of calcium tends to deposit the calcium inside the steel pipe until the pipe may actually be closed by the deposit. This condition does not exit when copper pipe is used. Copper is available as type L pipe or as type K tubing. The latter requires fewer fittings and may be bent instead of using fittings. To reduce "water hammer" in a water system, there are usually air chambers installed at ends of water lines. These air chambers provide a contained amount of air that is compressed whenever a water surge occurs, thus eliminating the noise of the change.

Figure 14-1 shows a typical installation in a multistory residence. The vent piping is installed to convey any gases generated by the sewerage waste to the open air. The traps at each fixture have a return bend that is continuously filled with water as the fixture is used, thus creating a water barrier to prevent sewer gas from entering the house through the sewer line. Vents from various fixtures are

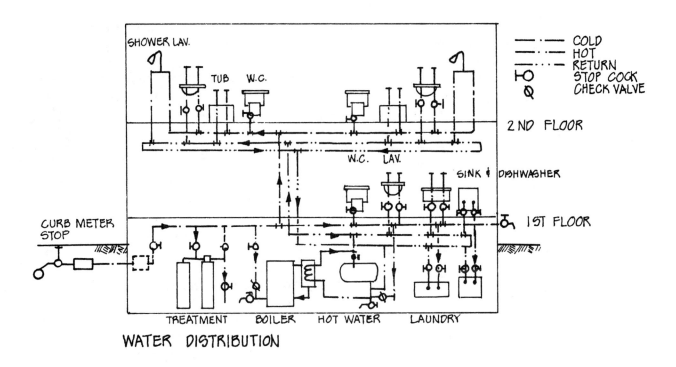

WATER DISTRIBUTION

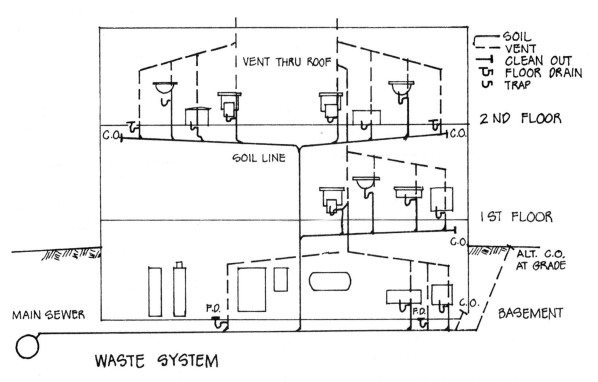

WASTE SYSTEM

**Figure 14-1** Multistory plumbing installation.

generally connected together so that only one stack protrudes through the roof. Cleanouts should be provided at ends of sewer lines and at fixtures where waste may clog, so that these lines may be cleared without dismantling the piping system.

**14.2 FINISH PLUMBING**   Installation of the fixtures and their controls is the primary subject of finish plumbing. Fixtures include tubs, lavatories, water closets, sinks, laundry trays, and similar parts which are to be connected to water and sewer piping. The controls are a variety of faucets, valves, volume and heat controls, and other fittings, usually chrome plated, and generally termed "brass," as this material is often the base metal used in the manufacture of the parts.

Tubs are manufactured of cast iron with a fired enamel finish or may be of pressed steel with a painted finish. Entire tub and shower units are now available in a variety of shapes and sizes. Lavatories are usually of cast ceramic (china) or may be of cast iron or pressed steel. Water closets are almost universally made of cast ceramic with a fired-on glaze finish and are formed with an integral trap. Kitchen sinks are available in a wide range of materials, sizes, and shapes but are most often of enameled cast iron, enameled pressed steel, or stainless steel with a natural finish. Laundry trays may be fabricated of pressed steel, cast iron, or various cement compounds. All the fixtures listed above, except water closets, require an exterior trap that connects to the sewer line.

Usually included, but often overlooked in an estimate, are the domestic water heater, sink-installed disposal unit, and proper connections for dishwashers, laundry dryers and washers, and other small but important items that require water, gas and/or sewer connection.

**14.3 SEPTIC TANK SYSTEMS**   Where municipal sewer systems do not exist, or in a few cases where connection may be difficult or impossible, the septic tank and the drain field may be a part of the plumbing installation requirements. The septic tank is a multicompartment container, usually buried in the ground, connected at the influent end to the house drain and at the effluent end to a drainage field or vertical cesspool. Waste material enters the tank with a considerable quantity of water, is disintegrated by bacterial action in the tank, the resulting sludge falls to the bottom of the tank, and the excess water enters the drain field. Occasional pumping of the sludge is required to maintain the efficiency of the tank.

Septic tanks are made of almost any liquid-retaining material but concrete, plastered concrete block, factory-fabricated steel, or plastic units are most common. The disposal field may be constructed of open-joint clay tile laid in a trench about 1'-6" wide by 3'-0" deep with pipe surrounded with crushed rock or pit-run gravel and covered with a layer of earth to grade level. Cesspools are vertical wells, lined with stone, open-joint masonry units, or wood cribbing of various types to allow the liquid to percolate into the surrounding earth. Adequate tests should be made on the earth to determine the porosity so that the estimated water flow may be accommodated. If a number of drain lines are used, a switch box adjacent to the tank may be required to allow some portions of the system to be used while other portions dry out (Figure 14-2).

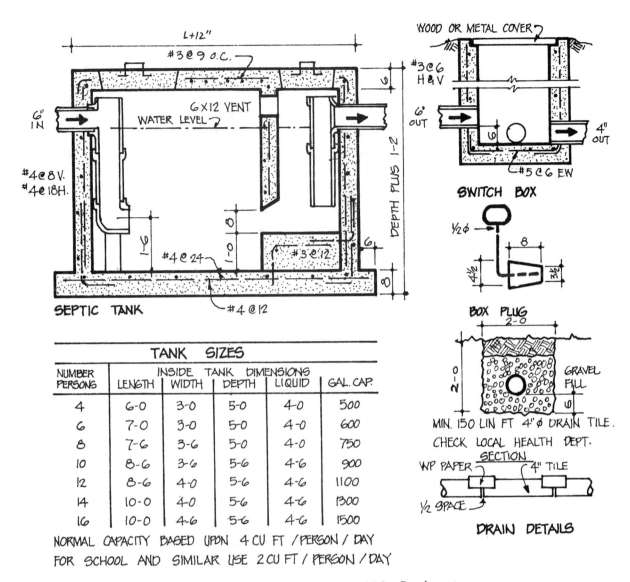

**Figure 14-2** Septic system.

| TANK SIZES | | | | | |
|---|---|---|---|---|---|
| NUMBER PERSONS | INSIDE TANK DIMENSIONS | | | | |
| | LENGTH | WIDTH | DEPTH | LIQUID | GAL. CAP. |
| 4 | 6-0 | 3-0 | 5-0 | 4-0 | 500 |
| 6 | 7-0 | 3-0 | 5-0 | 4-0 | 600 |
| 8 | 7-6 | 3-6 | 5-0 | 4-0 | 750 |
| 10 | 8-6 | 3-6 | 5-6 | 4-6 | 900 |
| 12 | 8-6 | 4-0 | 5-6 | 4-6 | 1100 |
| 14 | 10-0 | 4-0 | 5-6 | 4-6 | 1300 |
| 16 | 10-0 | 4-6 | 5-6 | 4-6 | 1500 |

NORMAL CAPACITY BASED UPON 4 CU FT / PERSON / DAY
FOR SCHOOL AND SIMILAR USE 2 CU FT / PERSON / DAY

## 14.4 PLUMBING COSTS

Because of the great number of fittings, sizes, and types of pipe, quality of fixtures, and the difference in labor required to install the plumbing, the average general contractor or self-builder is not in a position to estimate accurately the cost of plumbing, air-conditioning, or electrical work. In actual contracting practice these portions of the total work are estimated by the appropriate subcontractor. Although this procedure is common for normal projects, there are also occasions when in-place costs are valuable as well as unit costs for various quantities of material.

To obtain a reasonably fair cost figure we must first consider the similarity of fixtures. Tubs, lavatories, water closets, and kitchen sinks require approximately the same rough and finish plumbing work, so may be estimated at about $500 to $600 each in place and require about 5 hr of rough plumbing work and an average of about 3 hr to finish. If cabinets or ''cultured marble'' tops or counters are used to enclose lavatories or sinks, add the cost of the cabinetwork and top as necessary. Rough-in waste and vent per fixture averages about $300. Laundry trays of cement compounds, double compartment 24″ × 48″, cost about $500. Most residential water closets are floor mounted, but if wall-hung closets are used, add about $250 to $300. This figure should include supply and installation of required fixture hanger or "chair." Gas-fired 40-gal domestic water heaters cost about $350 in place; oil-fired glass-lined water heaters cost about $600 to $650. Most sewer and water lines are "stubbed out" about 5′-0″ beyond the building line, so sewer and water lines beyond this point must be figured in addition. Do not forget a gas line or black iron pipe if water heater, kitchen range, or space heating is gas-fired.

## 14.5 ROUGH SEWER COSTS

Cast iron pipe is most often used for sewer lines and is available in lengths of 5′ and 10′ and in many sizes. Extra-heavy-weight pipe 5′ long and 4″ in diameter costs about $5.50 per lineal foot, while 6″ extra-heavy pipe costs about $9.00 per lineal foot. This type of pipe may be installed by a three-man crew at about 50 lin ft per day or at an in-place cost of approximately $18 per lineal foot. Each 5′ length requires a lead-and-oakum joint seal which requires about 3 lb of lead and ¼ lb of oakum per joint. Lead currently sells for about $1.25 per pound and oakum sells for about $1.50 per pound.

Vitrified clay sewer pipe is also used outside buildings, is available 4″ to 36″ in diameter, in 48″ and 60″ lengths, with a variety of joining types. Material cost for the 4″ size is about $1.50 per lineal foot and about $2.25 per lineal foot for 6″ diameter. Joints may be made with metal slip joints or cement and require about 4 hr per 100 lin ft utilizing a three-man crew at an in-place cost of about $5.50 per lineal foot. Some areas now permit fiberglass plastic pipe to be used for "in-house" drainage lines or polyvinyl chloride plastic in sizes from ¼″ to 6″ diameter. Plastic soil line 4″ in diameter schedule 40 in 10′-0″ lengths costs about $2.85 per lineal foot and should require a two-man plumber team about 1 hr per 10′ length at an in-place cost of about $15 per lineal foot including couplings, hangers, and other accessories. Cleanouts for 4″ pipe cost about $100 in place. Soil bends and other similar fittings may be estimated at about 1½ to 3 times the cost of straight pipe. When estimating the cost of buried pipe, be sure to include the cost of excavation and back-fill as a separate and additional item.

## 14.6 ROUGH WATER COSTS

Water pipe is available in galvanized steel, copper, plastic, and stainless steel or glass. The latter two are normally not used in residential work but are required in situations where distilled water or chemicals are used. Galvanized steel pipe with threaded cast steel fittings is the most usual type of water pipe. This type has problems, however, when water contains a high calcium count (termed "hard water"). Over a period of years the calcium is deposited on the inside of the pipe until the water passage is considerably reduced or even closed. Copper pipe or tubing with brazed brass or bronze fittings eliminates this problem and is now much more popular despite periodic shortages of material and higher costs. Polyvinyl chloride (PVC) and similar plastics are also available for water piping, but special care should be exercised if plastic pipe is to be used for hot-water lines. Some plastics and some types of plastic pipe tend to fail when exposed to heat greater than 100°F. Gas lines should always be installed with black iron pipe, as the gas chemicals affect the zinc of the galvanizing process and could cause clogging of pipe or gas jets. Some lead pipe is still used for toilet bends but is not normally used for water piping.

Galvanized steel pipe is available from $\frac{1}{4}$" to 3" in diameter. Fittings and connections are made with threaded joints and joint sealer. Installation averages about 65 to 75 lin ft per day and costs about $4.25 per lineal foot for $\frac{1}{4}$" pipe to $150.00 for 3" pipe in place. Copper pipe is available in hard-tempered 20' lengths or in soft-tempered coils, from $\frac{1}{4}$" to 2" in type K and to 6" in type L. Fittings are installed with molten solder. Installation of $\frac{1}{4}$" copper pipe costs about $4.10 per lineal foot installed at about 50 to 60 lin ft per day. One-inch copper pipe in place costs about $8.50 per lineal foot. PVC plastic pipe is available from $\frac{1}{4}$" to 6" in diameter, with $\frac{3}{4}$" to $1\frac{1}{2}$" most used. High-strength plastic pipe requires about 1hr for each 5 lin ft at a cost of about $11.50 per lineal foot. ABS (acrylonitrile–butadiene–styrene) plastic pipe for drain, waste, and vent (DWV) size $1\frac{1}{2}$" costs about $8 per lineal foot in place. Polyvinyl dichloride (PVDC) plastic pipe is the only one manufactured for hot-water lines and costs about the same as for other types of plastic.

Miscellaneous water fittings include pressure valves at approximately $40 each for 1" pipe, check valves at $40 each for 1" pipe, and various valves in 1" size from $20 to $50 each. Installation time required for valves and similar work is about $\frac{3}{4}$ to $1\frac{1}{2}$ hr per unit. Final testing of water lines as well as sewer lines is usually done by plugging all outlets, then filling the systems with water or gas, and checking for leakage with a pressure gauge installed in an opening. Such testing may be estimated at about 2 hr per fixture.

## 14.7 FINISH PLUMBING COSTS

The fixtures are the principal items included in the finish plumbing work, together with their installation, the "brass" faucets and other fittings, and the final connections to water and sewer. Cast iron tubs, 60" × 30", with porcelein fired-on finish cost about $550 to $600 each in place, while pressed steel enameled tubs of the same size cost about $250 each in place. Complete fiberglass or other plastic preformed units cost about $300. Tubs require about $2\frac{1}{2}$ hr each for a two-man team to install and about 1 hr for one plumber to install the faucets and other fittings. If shower mixing valves are used, they require about another hour of installation time and cost about $130 to $150 in place.

Water closets, two-piece, floor-mounted, white ceramic (china) cost approximately $125 to $150 and require about 1½ hr to install. The only finish material for a water closet may be a ⅜″ stop valve on the water line installed at a cost of about $40 each. Lavatories may be vitreous china, enameled steel, stainless steel, or molded plastic. The cost of these fixtures ranges from about $125 to about $175. Chrome-plated faucet sets for lavatories cost $35 to $40 and require about 1 hr of a plumber's time for installation. Kitchen sinks of enameled steel, 24″ × 21″ single bowl, cost about $200 in place, with installation by a two-man team at about 2½ to 3 hr per unit. A stainless steel single bowl of the same size costs about $50 to $60 more. Faucet sets for sinks cost about $50 in place and require about ¾ to 1 hr for one plumber to install.

## 14.8 SEPTIC TANK COSTS

Septic tanks for residential use should be calculated at about 50 to 60 gal per person per day with a minimum size of 1000 gal. This means that the actual tank will be approximately 5′-0″ × 5′-0″ × 10′-0″ and may be of concrete or coated steel construction. A septic tank unit, purchased factory-fabricated, costs about $300 for the 1000-gal size. In addition to the cost of the tank the estimator must include the cost of excavation for the tank, back-fill after the tank has been installed, and perhaps the cost of renting a crane to place the tank in the excavation. Cast-in-place concrete septic tanks should be estimated as forms, reinforcing, concrete, and miscellaneous parts as well as for excavation and back-fill.

Drainage fields of clay tile buried in trenches with the pipe surrounded with crushed rock or gravel require about 200 lin ft of 4″ pipe laid with ½″ open joints. Excavation for the disposal field by trenching machine costs about $1 per lineal foot and about $3 per lineal foot for back-fill and compaction. Gravel fill costs about $7 per cubic yard and usually requires about 1 hr for each 20 cu yd in place at about 2¼ cu ft per lineal foot of drain or about 1 cu yd of gravel for each 12 lin ft of drain line. Switch boxes may be required to change disposal field use and average about $100 each fabricated in place with concrete.

## 14.9 SITE PLUMBING WORK

Sewer, water, and gas lines required to connect the "house system" to the public system in the street are usually included as plumbing work and should be thoroughly considered. Some specifications require all excavation and back-fill for this piping to be included as plumbing work, while in actual practice this work may be more easily done when machines used for trenching of footings and foundation work are still available on site, so may not be plumbing work. On large projects various types of corrugated steel drain pipe and culvert material for storm drains, as well as asbestos-cement pipe, may be used by specialized contractors. These systems are not normally required in residential work so are not considered further in this text. Utility companies may also install service piping from the street line to the meter or building location. For the most part this type of installation is similar to trenching and pipe laying as indicated in other parts of this text.

## PROBLEMS

**14-1.** Estimate the cost for installing 6″ cast iron sewer piping, 1″ copper water line, and 1″ black iron gas pipe from the street center-line 40′-0″ from the property line to the residence for which plans are provided on pages 194-197.

**14-2.** What is the preliminary cost estimate for the plumbing fixtures in these same house plans?

**14-3.** Assume that a septic tank and 300 lin ft of drainage pipe with one switch box are required. What is the approximate in-place cost?

# chapter

# 15

# Heating,
# Ventilating,
# and
# Air Conditioning

The title of this chapter says it all. Discussed are various heating systems, ventilation, and air conditioning, generally referred to as HVAC. The range is tremendous. Heating technically includes every type of heat source from wood-burning stoves and fireplaces through the latest coal-, oil-, or gas-fired units. Ventilation is concerned primarily with commercial and industrial dissipation of foul air and gases, although some ventilation is normal in kitchens and bathrooms. Modern homes generally combine heating and ventilating with cooling, so air conditioning is a large, and growing, portion of the total work. This work, like that connected with plumbing and electrical, is highly specialized and is usually installed by subtrades.

Air-conditioning ducts, required piping or electrical work, and control locations are usually shown by symbols. This again means that the estimator may have to actually design the system before it can be estimated. Most times the location of the main unit is shown as an oblong inside or outside the building, a heavy line indicating outlet grilles with an arrow and cubic feet per minute (CFM) shown, return air grille, and thermostat location. The size of ducts, their shape, location in the structure, and other details are left to the contractor. Without specialized knowledge or a complete set of HVAC drawings, anyone except a specialized estimator may be simply guessing at in-place costs. For all practical purposes we will disregard wood-burning equipment in this text, both for its limited use and for the shortage of the fuel, although good authority claims that wood, a reproducible energy source, will be providing up to one-tenth of the world's energy by the year 2000. It is also estimated that oil and gas supplies will be used up by about 2025 and coal supplies will last only about 125 years.

## 15.1 HEATING SYSTEMS

Two general types of heating systems need to be considered: a system of piping utilizing hot water or steam, and a central heating generator plus ducts that conduct the heated air to the location where it is required. In either case the fuel used may be coal, oil, or gas. Storage of coal or oil may be a problem with some equipment, especially oil, where a buried tank of 250 to 500 gal may be necessary. Coal is normally stored in a bin or room adjacent to the heating unit. A rather complicated formula has been devised by the American Society of Heating, Refrigerating and Air-Conditioning Engineers (ASHRAE) to determine the heat loss, which in turn determines the type and amount of heat required for any structure.

## 15.2 HOT-WATER HEATING

Hot-water heating is one of the older types and requires a central water heater or boiler, extensive piping, and an arrangement of radiators to distribute the heat derived from the hot water. Boilers may be water-tube, which means that the fire surrounds tubes that carry the water, or may be fire-tube, in which the fire passes through tubes surrounded by water to be heated. Boilers may be of cast iron connected sections or of steel and may be contained in many different shapes.

After heating, the water in the boiler is moved into the piping system by temperature action or a small pump. Two major systems are in use, the one-pipe system, in which the hot water continuously passes from one radiation unit to the next, or the two-pipe system, in which the cooled water returns to the boiler by a separate pipe (Figure 15–1). Steam systems are similar to hot-water systems except that the steam tends to condense back into water and is drained off to the boiler. Radiation units may be cast iron vertical units or may be horizontal tube types equipped with metal fins as a type of baseboard unit.

Hot water may also be used in a *radiant* heating system. In this system small pipe, usually of copper and about ⅜″ or ½″ in diameter, is embedded in the concrete floor slab in a series of parallel runs spaced about 6″ to 8″ apart. Rooms may have separate systems with separate controls. Careful installation is necessary to assure a uniform heating surface. The hot water is supplied by a central "boiler" which is usually gas-fired. Radiant heating keeps the floor warm, rarely produces drafts, and tends to warm all persons, furniture, and accessories by direct radiation. However, if a pipe breaks or corrodes, it may be difficult to repair, as it is buried in concrete and may require removal of the entire floor in order to find a leak.

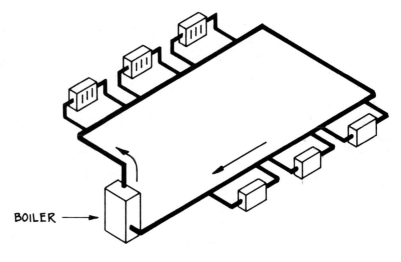

ONE-PIPE SYSTEM

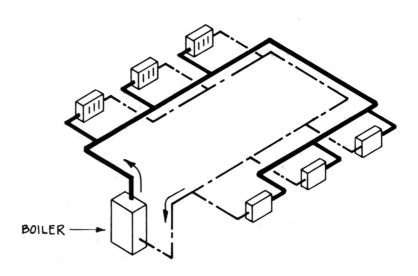

TWO-PIPE SYSTEM

**Figure 15-1**  Hot-water heat pipe system.

**15.3 HOT-AIR HEATING**

Hot-air heating systems may be of two major types. First is a rather large furnace consisting of a cast iron fire box surrounded by an air chamber, a system of ducts to distribute the heated air, and a return duct for "used" or fresh air (Figure 15–2). Fuel may be wood, coal, oil, or gas. The second system consists of a smaller unit, usually gas-fired, which heats the air and distributes it through a system of duct work with the assistance of a fan, and obtains return air or fresh air by duct work. Either of these systems, except perhaps when wood-fired, may be automatically controlled for fuel supply as well as temperature. The latter type is often a part of an air-conditioning installation.

Duct work includes all the "pipes" from the heating plant to

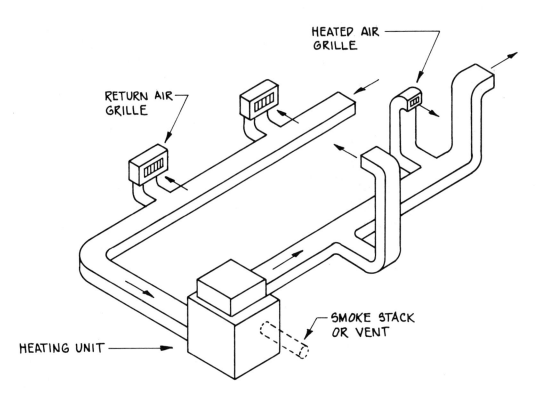

**Figure 15-2** Hot-air ducting.

the various room outlets. Several systems use large plenum spaces to conduct the initially heated air from the heating unit and then is further distributed with smaller ducts to the required locations. Duct systems are usually installed in attic spaces but may also be installed beneath the floor in crawl spaces. Air outlets should be adequate in sizes to provide continuous comfort in the space and the best location in most cases is beneath a window, to offset any downdraft from the cold air flowing down the window pane. Sizing of ducts for heating should also be sized for air conditioning, even if the air-conditioning system is not immediately installed.

**15.4 AIR CONDITIONING**

Air conditioning is an extension of the heating system to include cooling when exterior temperatures become excessive. The simplest system is a self-contained unit, using either water evaporation for cooling the air or refrigeration coils. Units may be installed in windows, in attics, or other convenient locations without duct work. When water evaporation units (colloquially termed "swamp coolers") are used, a small-pipe water supply is necessary and a unit tray for drainage must be installed. For refrigeration-type units the required condenser must be outside the building and a drip pan provided for condensate.

Another system commonly used for residential installations requires that the condenser and the compressor unit be installed on a concrete pad outside the building with the heating–cooling–blower section inside the building. The air-conditioned air is then distributed throughout the building by a duct system and controlled with duct dampers and adjustable grills or registers. This entire operation is usually automatically controlled by thermostats and automatic switching. For commercial installations the cooling system is often a water tower which replaces the small condenser used in residential work but is a major structure and requires considerable construction and piping work.

## 15.5 AIR-CONDITIONING CALCULATIONS

To arrive at any cost figures for heating or air conditioning we must first try to determine the requirements obtained by calculating the possible heat loss or gain. For absolute accuracy each wall, window, door opening, or other surface or opening should be figured separately and the total arrived at. This is a rather long and complicated job, but some shortcuts are available. First find the cubic volume of the room or space, then multiply this with a loss factor of about 6.0 for exposure of walls to exterior on all sides to a low of about 3.5 for exposure to common heated surfaces, walls, floors or ceilings. If the expected low temperature is approximately 0°F, multiply by 1.00; if higher than 0°F, multiply by 0.50; or as low as −20°F, multiply by 1.25. By multiplying the volume by the factors above an estimate of BTUH (British Thermal Units per Hour) may be obtained that will maintain the required temperature in the room.

Knowing the requirements to maintain temperature, an estimate is then made for equipment to provide adequate heating or cooling. Using cast iron radiators, in a steam or hot-water system, 1 sq ft of EDR (Equivalent Direct Radiation) equals about 150 BTUH for hot water and about 240 BTUH for steam. If convectors are used, 4 sq ft of EDR equals 1000 BTUH. For hot-air systems the heat is provided on CFM (Cubic Feet per Minute) derived from the following formula:

$$CFM = \frac{BTUH}{1.08 \times \text{temperature difference between supply air and room air}}$$

Many charts are available to calculate heat loss (or temperature replacement) and requirements are usually already indicated on construction drawings. Typical requirements for residential air conditioning are calculated at 20 Btu per square foot or 600 sq ft per ton of air conditioning.

## 15.6 VENTILATION

When air conditioning is used, the requirement for additional ventilation may often be waived if the per-minute air change is adequate. In a great many conditions, and as a requirement of some codes, additional ventilation is required in bathrooms and kitchens. In baths this ventilation is often considered to be provided by operating windows, but in enclosed baths where no windows are provided there must be a fan connected to the light switch so that ventilation is automatically supplied when the room is lighted and in use. Kitchen fans are usually placed at range locations or in a range hood over ranges to carry away cooking odors. Separate fans cost about $75 per unit in place. Where a combination bath light, fan, and infrared heating fixture is installed, figure about $110 to $125 in place. For a kitchen range hood 30″ wide, two-speed, vented, estimate about $275 in place.

## 15.7 HEATING AND AIR-CONDITIONING COSTS

Heating and air-conditioning costs vary so much that about the best method for determining these costs is to rely on a contractor specializing in these fields. A boiler system with hot water or steam, used in the eastern and northern parts of the United States, is entirely different from gas-fired space heating using warmed air. Some of the estimated costs included in this chapter may surprise people, but heating and air conditioning is currently a very large portion of total construction costs.

For a hot-water piped system with cast iron radiators, the boiler costs about $4000 for 400 MBTU in place, radiators about $20 each, and piping of galvanized steel about $6 per lineal foot, with controls costing from a minimum of $100 to perhaps as much as $1000, depending on the system. Baseboard units for hot-water heating average about $40 per lineal foot. Radiant heating in the concrete floor slab using copper piping should be calculated at about 40 to 50 Btu per square foot of floor area and costs about $3.25 per square foot of floor area.

Air-conditioning systems for residences, using a self-contained package, cost about $10,000 for 30-ton capacity. In general, the cost for heating and air conditioning for residential use may be estimated at about 4% of total cost of construction at the low end and about 8% at the high end. Estimating of mechanical systems is not for the amateur or beginner, so in this phase at least consult with a knowledgeable specialty contractor in these fields.

## PROBLEMS

**15-1.** Estimate the cost of air conditioning for the house plans included in this book, pages 194-197, using the formula given and your local range of inside and exterior temperatures.

**15-2.** Estimate the cost of range hood and bath ventilation for the same house plans.

**15-3.** Check your own living quarters to see if adequate ventilation is provided for bath and kitchen.

**15-4.** Check your own living quarters to see if an adequate heating facility is provided.

# chapter

# 16

# Electrical Work

Electrical work, like mechanical work, is a specialized trade and is estimated and installed by subcontractors. Here again there is both rough and finish work for electrical installation. Rough electrical work includes installation of service equipment, conduit and wiring, boxes for switches and outlets, and similar work that is mostly hidden by wall or ceiling finishing material. The finish work includes the installation of switches, convenience outlets, lighting fixtures, connection of these parts to the wiring system, and perhaps the final connection to motors, fans, and similar equipment. Low-voltage and communication systems are installed in approximately the same manner.

In a manner similar to plumbing work, residential electrical systems are usually shown only by location of symbols on the floor plans, representing fixtures, switches, and outlets. Electrical estimators must then actually design the system, balance the panel loads, and conclude how all of the concealed work may be economically placed. Conduit and wire sizes are regulated by codes as well as by some local ordinances. Estimating of electrical work is done exactly like any other type of construction, by a quantity survey of materials and labor, plus costs for overhead and profit. Also, like mechanical work, a cost-in-place per unit may be an easy way to estimate roughly the final cost of installation.

## 16.1 ELECTRICAL SERVICE

Voltage service for residential use is normally limited to 110/120 volts or to 220/240 volts, single phase. Lighting and most convenience outlets are most conveniently wired for 110/120 volts, as this is the rated voltage for portable lamps and similar furnishings.

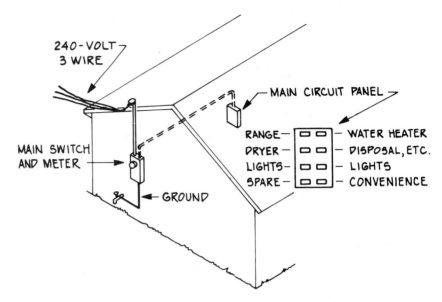

**Figure 16-1** Three-wire system voltage split.

Ranges, dryers, and other high-usage heavy-duty equipment require 220/240 voltage. Service from the utility system requires only two feed wires for 110/120-voltage service but requires three wires for 220/240 volts. If the three-wire system is used, the voltage may be split as shown in Figure 16–3 to provide both 110/120 and 220/240 supply. Most utility service is provided overhead, as shown in Figure 16–1, through rigid steel conduit and a weatherproof "pot head," then to the meter and safety switch. Grounding of the system is accomplished by proper connection to a water pipe or other approved ground, *not to a gas pipe*. The safety switch allows all electricity to be disconnected at one location in case of emergency and may be fused to prevent overloading of the entire system.

## 16.2 ROUGH ELECTRICAL WORK

Almost without exception the conduit, wiring, and accessory boxes are installed while the rough framing of the building is still exposed and before any finish wall or ceiling material is applied. Electrical wiring may be of non-metallic-sheathed cable (Romex) which is composed of two to four insulated wires grouped into one unit and further protected with an additional covering, usually black. This material does not require additional conduit or other encasement and is secured to outlet boxes with clips. An additional grounding wire, often bare wire, is needed and must be attached to boxes with a screw connection. Unfortunately, this system is not allowed for residential construction by many city codes and not at all for commercial work.

The second system is used for all types of construction and consists of hollow thin-wall metal tubing (EMT) through which the required number and type of wires are pulled after the conduit is installed. The tubing is secured to boxes with clamps and threaded

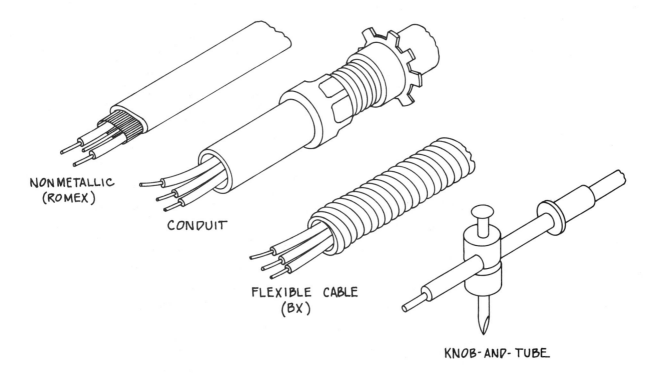

NON METALLIC
(ROMEX)

CONDUIT

FLEXIBLE CABLE
(BX)

KNOB-AND-TUBE

**Figure 16-2**  Typical electrical diagram.

fittings. This system provides an automatic continuous grounding path if properly installed. Wire is normally of copper enclosed with colored insulation (white, red, black, blue, or green), which facilitates proper connections. The size of wire and type of insulation are specified by code, as is the size of the conduit to accommodate the number of wires required in the circuit. Rigid pipe is also used as conduit in some locations, primarily as service from the initial service connection to meter, safety switch, or distribution panels.

Metallic cable is composed of a number of insulated wires enclosed in a spirally fabricated steel core. This material is generally used where flexibility is a requirement, where conduit cannot be bent to a short-enough radius on a turn, or for final connections to equipment or machinery. In older buildings a "knob-and-tube" system (Figure 16–2) may still be found, although this method is now obsolete and is banned in most places.

Wire used in residential work in most cases is #14 size minimum with #12 more adequate. Location and usage determine the type of insulation cover required and may be of several varieties. For ranges and similar heavy-draw units #10 or #8 wire is usually required, and powered equipment may demand wire as large as #2/0. Outlet boxes required for fixtures are generally of galvanized steel, 2″ × 4″ or 4″ × 4″, or may be 4″ octagonal for intersecting or joining conduit. Junction boxes are usually referred to as "J boxes." Plastic boxes are also available and are allowed in some locations. Plastic boxes have side tabs or holes for attachment to studs and are the same nominal size as metal boxes.

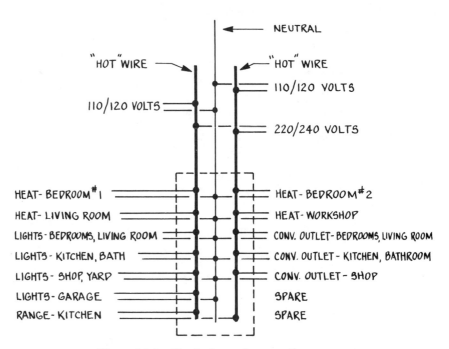

NEUTRAL

"HOT" WIRE →

"HOT" WIRE

110/120 VOLTS

110/120 VOLTS

220/240 VOLTS

HEAT- BEDROOM #1 — HEAT- BEDROOM #2
HEAT- LIVING ROOM — HEAT- WORKSHOP
LIGHTS- BEDROOMS, LIVING ROOM — CONV. OUTLET- BEDROOMS, LIVING ROOM
LIGHTS - KITCHEN, BATH — CONV. OUTLET - KITCHEN, BATHROOM
LIGHTS - SHOP, YARD — CONV. OUTLET - SHOP
LIGHTS- GARAGE — SPARE
RANGE- KITCHEN — SPARE

**Figure 16-3**  Typical panel connections.

**16.3 CONTROL EQUIPMENT**

A certain number of switches, ciruit breakers, starters, and similar control equipment are needed to operate the various electrical circuits properly. In residential work a toggle switch, used singly or in gangs, is the most common control device to turn electricity on or off. To control, or rather to protect, circuits from overloads and the danger of fire, fuse boxes or circuit-breaker panels are used. Fuses may be cartridge type or screw-in glass type. Either type contains a thin metal strip rated to definite load capacity which will melt on overload, thus disconnecting the circuit. Circuit breakers are miniature enclosed thermostatic switches that open automatically on overload. External switches on each breaker unit may be used to reactivate the breaker when the overload has been removed. Fuses must be replaced when the metal strip burns through. Starters are heavy-duty enclosed switches, properly fused, to protect machinery or equipment overload.

**16.4 FINISHING ELECTRICAL WORK**

Selection and installation of lighting fixtures and lamps, installation of switches and convenience outlets, and the installation of miscellaneous small parts constitutes the finishing electrical work. Lighting fixtures probably create the most problems, as there are so many different designs and types available from so many different manufacturers. Most fixtures are supplied with a bracket or plate that is secured to the outlet box, the fixture wires and circuit wires are connected together, and the fixture cover conceals the connection. Lamps for fixtures may be incandescent or fluorescent and may or may not be required as a part of the electrical work.

Convenience outlets should be wired to a different circuit than lighting. This is of particular importance in kitchens, garages, shop areas, and similar locations where lighting is still needed when equip-

Table 16-1  Conduit wire capacity

| Wire size | | | | | Maximum number of conductors in conduit or tubing | | | | | | | |
|---|---|---|---|---|---|---|---|---|---|---|---|---|
| | ½ | ¾ | 1 | 1¼ | 1½ | 2 | 2½ | 3 | 3½ | 4 | 5 | 6 |
| 18 | 7 | 12 | 20 | 35 | 49 | 80 | 115 | 176 | — | — | — | — |
| 16 | 6 | 10 | 17 | 30 | 41 | 68 | 98 | 150 | — | — | — | — |
| 14 | 4 | 6 | 10 | 18 | 25 | 41 | 58 | 90 | 121 | 155 | — | — |
| 12 | 3 | 5 | 8 | 15 | 21 | 34 | 50 | 76 | 103 | 132 | 208 | — |
| 10 | 1 | 4 | 7 | 13 | 17 | 29 | 41 | 84 | 86 | 110 | 173 | — |
| 8 | 1 | 3 | 4 | 7 | 10 | 17 | 25 | 38 | 52 | 67 | 105 | 152 |
| 6 | 1 | 1 | 3 | 4 | 6 | 10 | 15 | 23 | 32 | 41 | 64 | 93 |
| 4 | 1 | 1 | 1 | 3 | 5 | 8 | 12 | 17 | 24 | 31 | 49 | 72 |
| 3 | — | 1 | 1 | 3 | 4 | 7 | 10 | 16 | 21 | 28 | 44 | 63 |
| 2 | — | 1 | 1 | 3 | 3 | 6 | 9 | 14 | 19 | 24 | 38 | 55 |
| 1 | — | 1 | 1 | 1 | 3 | 4 | 7 | 10 | 14 | 18 | 29 | 42 |
| 0 | — | — | 1 | 1 | 2 | 4 | 6 | 9 | 12 | 16 | 25 | 37 |
| 00 | — | — | 1 | 1 | 1 | 3 | 5 | 8 | 14 | 14 | 32 | 32 |
| 000 | — | — | 1 | 1 | 1 | 3 | 4 | 7 | 9 | 12 | 19 | 27 |

ment may cause the protective device to disconnect the service. In rooms where no fixed light fixture is installed, the electrical code requires that at least one convenience outlet be controlled by a switch. In this case one side of the duplex outlet should be switch-controlled, with the remaining side "hot" all the time. Dimmer controls may be advantageous instead of toggle switches in some locations. Where switches are located at stairs or other positions where the controlled light is not easily seen, a pilot light may be used in conjunction with the switch to indicate the condition of the light. Properly installed toggle switches should have the switch handle in a *down* position when in the "off" condition. If a switch accidentally break when the electricity is "on," the switch may fall, thus disrupting the circuit.

## 16.5 LOW-VOLTAGE ELECTRICAL WORK

Low-voltage installation is rather limited in most electrical work and in residential construction usually includes only door bells or chimes and possibly some lighting controls. Voltage is considered "low voltage" when it is transformed down to 6 to 24 volts from the normal 110/120 volts. This is easily done with a compact transformer unit not often larger than $4'' \times 4'' \times 6''$ installed in a closet or similar location. Pushbuttons for door bells are connected to the bell or chimes with either bare wire or lightly insulated wire of about 20 gauge which does not require conduit or other protection. If low voltage is used to control lighting, the low-voltage actuates another switch, which in turn operates the line voltage to the light fixture.

## 16.6 ELECTRICAL COSTS

Most service connections consist of a length of rigid galvanized steel pipe capped with a service "pot head" to accommodate three wires, and terminates at the lower end with a connection to a fused safety switch and meter. The rigid conduit, $1\frac{1}{2}''$ in diameter, costs about $5 per lineal foot in place plus about $30 for the terminal pot head. The safety switch for most residential work will be three-pole of 240-

volt capacity, rated at 100 amperes, although this rating may be greater depending on the house load. Such a panel costs about $180 in place. The meter ring is usually a part of the electrical installation and costs about $50. The meter will be supplied by the utility company. Main service from the safety switch to an interior circuit-breaker panel is usually 1½ ″ conduit or EMT tubing with a three-wire service not smaller than #8 wire at an in-place cost of about $3.50 lineal foot for conduit, $5.00 per lineal foot for three wires, at a rate of installation of about 15 lin ft per hour.

Main circuit-breaker panel size depends on the number of circuits to be used, with most circuits being rated at 15 amperes, but with some heater, range, and similar equipment receiving 20-amperes breakers or possibly having two circuits ganged together. These panels range from about $75 for 4 circuits to $175 for 12 circuits. Nonmetallic-sheathed cable (Romex) with two #14 wires plus a ground wire costs about $7 per lineal foot in place, while metallic tubing (EMT) costs about $2 per lineal foot for the 1″ size plus $2 per lineal foot for each wire pulled in the conduit. Tubing of 1 ″ size should be installed at about 15 lin ft per hour, while Romex is installed at about 30 lin ft per hour. Grounding connection costs $30 each and each steel box for switches, convenience outlets, and lighting fixtures costs about $10 each in place. Installation of these boxes is at an average of three per hour. Switches for control of lighting and convenience outlets each cost about $18 complete, and range and heater circuits average about $60 each.

Lighting fixtures may be completely specified by manufacturer number, size, shape, wattage, and so on, but this procedure is rather rare in residential construction. In place of this detailed list there is often an "allowance" figure given to the electrical contractor. As explained in connection with finishing hardware (Chapter 8) this allowance is for materials (fixtures) only and is used as a base for the purchase of lighting fixtures. In practice, the owner, architect, and contractor go to a lighting fixture supply house and select what is needed; if the total fixture costs exceeds the allowance, the owner pays the additional. On the other hand, if the allowance is greater than the fixture cost, the contractor pays the owner the difference. The cost of fixtures runs from about 10 to 12% of a total project cost. Installation time, not included as a part of "allowance," averages about 1 to 1½ hr per fixture for one electrician. Lamps are not always included in the electrical contract but probably should be to provide a truly finished job. Incandescent bulbs in the range 50 to 150 watts cost about $2 each in place, while 48 ″ 40-watt fluorescent tubes average about $5 each in place.

## 16.7 MISCELLANEOUS ELECTRICAL WORK

Several minor items of electrical work may or may not be required in the average residence. Almost every apartment or residence requires some sort of door bell or chimes, which cost about $75 for transformer, pushbutton, bell or chimes, and installation. In-house intercom systems using telephones cost from a low of about $75 to

$1000 or more for extensive or remote stations. Apartment call-in and door release systems range from $75 to about $150 each. Fire alarm and smoke detector systems average about $100 per unit in place.

## 16.8 ELECTRICAL HEATING

Electrical heating for residential projects is practically limited to two major types: resistance wiring buried in ceiling plaster and unit baseboard heaters. Electrical heating poses an additional load, so consideration must be made for increased wire size throughout and for control equipment that may normally be overlooked. Availability of adequate electrical supply at economical rates is also a major item for consideration. Amount of wiring or baseboard heating needs to be carefully calculated depending on the location in the country, inside and outside temperatures, type of construction, and various other considerations. Once an inadequate electrical system is installed it is very difficult to change, as most of the system is concealed beneath other surfaces and any change involves major remodeling.

Electrical heating for resistance wiring (cable) embedded in plaster costs about $6.00 to $6.50 per square foot plus about $50 for controls. Remember to figure the cost of plaster in addition. Baseboard units are rated by wattage per unit and range from 36" to 96" long, with or without a fan, from 500 watts to 1500 watts per unit. Cost for baseboard units range from about $65 to $125 or about $100 per kilowatt (1000 watts). Heater circuits from the baseboard to the circuit-breaker panel cost about $30 to $40. Wall heaters are often used to supplement other types of heat in bathrooms and similar locations. These are permanently installed, individual unit, recessed, direct radiant heat, manually or thermostatically controlled, ranging from 1000 to 4000 watts. The cost is from about $100 to $200, depending on size, fan, and controls. Another supplemental heating type is the ceiling infrared bulb and wall control. The fixture may be a reasonably priced screw outlet with toggle switch at a total cost of about $50 to $60 or with a timer switch at $25 additional. Be sure to check wire sizes to these heaters, as the load will often require a heavier wire than would normally be installed.

## PROBLEMS

**16-1.** Calculate the approximate total cost for electrical fixtures for the house plans on pages 194-197 if average cost per fixture cost per fixture is $46.

**16-2.** What is the total cost for all electrical work required by these house plans?

**16-3.** If radiant ceiling heating is used in the living room and each bedroom, how much will this cost?

chapter

# 17

# Commercial Estimating

Estimating commercial work, industrial projects, or any other type of construction is simply an extension of the methods suggested in this book. Almost every project has approximately the same general groups of work and by that token in actual practice the various trades are represented by subcontractors to the general or "prime" contractor. In each of these cases the general contractor has to determine what part of the entire project his own crew will do and then must obtain bids for the remainder of the work from a great variety of subcontractors.

## 17.1 THE SUBCONTRACTOR

Most subcontractors perform only a limited amount of the work required by the completed project. This is especially true of mechanical, electrical, air-conditioning, elevators, and similar trades. In some cases the general contractor may not do any of the actual construction work but may simply serve as a coordinator, in which case all the construction is done by subcontractors. The subcontractor then must make his own take-off of the work that he agrees with the general contractor to do, price the materials, figure the time costs, add overhead and his profit, and bid to the general contractor. This is usually done as a *quantity survey,* as very few projects are similar enough and costs for materials and/or labor may have changed from the "last job" if it is used as a guide. The subcontractors then submit an "estimate" (truly a firm figure) to the general contractor for consideration in putting his total bid together. On many projects there may be as many as a thousand subcontractors submitting estimates and in most cases there is only one (the lowest) that will be taken for each trade.

## 17.2 MANUFACTURER OR DISTRIBUTOR

Standing behind the subcontractor is the supplier of the material required by the subcontractor to complete his portion of the work. In most cases the subcontractor simply asks the price of a predetermined quantity of materials, but in some cases the distributor, or even perhaps the manufacturer, must make a take-off of materials and labor. This is especially true in a case where elevators or similar equipment is both manufacturered and installed by a "supplier." Again the quantity survey method is used, if only to separate the various types of material in order to price it. Cost of material, delivery, taxes, special packaging, and similar requirements are used, plus company overhead and anticipated profit. In a manner similar to the relationship between the subcontractor and the general contractor, the subcontractor may ask for material bids or "estimates" from a number of suppliers.

## 17.3 THE GENERAL CONTRACTOR

The general contractor is the final resting place for all the various bids or *estimates* pulled together by the manufacturers, distributors, and subcontractors. In addition to all of these quotations the general contractor makes a take-off for all the material, labor, and equipment that he will use to supply and install the work that his own crew will do on the project. In addition, he will juggle the various subcontract estimates to make what he may assume will be the winning construction bid to the owner. All of this again comes back to the principles of quantity survey methods.

## 17.4 NONRESIDENTIAL CONSTRUCTION

This book is intended as an introduction to the process of construction estimating and as such has been directed to the construction best known to the largest group of people—the residence. Regardless of the size of the project, the same principles may be used to estimate the cost of other construction. Commercial construction may have more types of subcontractors involved or may have more different materials included, but the process is the same—use "cost per square foot" for early preliminary "guesstimates," then make a series of current quantity surveys as the work progresses so that exact quantities may be determined. In buildings more than two stories high the additional cost for scaffolding or exterior elevators, cranes, on-site equipment of various type, and the higher cost of various insurance premiums must also be included. Quite often the matter of interest on borrowed money to finance the contractor's operation is not included, but most projects of a commercial nature require a longer period for construction and contractors borrow the money needed to process the work between owner's payments.

## 17.5 CONCLUDING REMARKS

Good, well-grounded, and accurate estimators are always in demand in almost every industry. Although this book deals only with building construction, the principles are the same for manufacturing the various materials, for operating a retail or wholesale business, or even for personal use in evaluating possible purchase of a product. In most cases a "method" for making a take-off is necessary. The estimator needs to have a regular way in which to approach the problem, then consider other aspects, and finally know the amount necessary for overhead of operation and the desired profit. For those readers who may want to estimate the costs for construction or remodeling of their own real estate, it is hoped that they will realize that any of the costs given in this book are those current at time of writing and are given so that serious students of estimating may have some figures by which they may be able to complete problems assigned to them. For those who are not in this situation the author suggests strongly that they contact subcontractors or suppliers for current prices.

# Glossary

**ABS:** Rigid plastic pipe compounded as acrylonitrile–butadiene–styrene.

**A/E:** Abbreviation often used for architect/engineer.

**Aggregate:** Rock, sand, or gravel used in the manufacture of concrete.

**Air chamber:** Short length of pipe at the end of water piping used to reduce air hammer.

**Alternating current:** An electric current that is purposely changed in direction of "flow" to facilitate distribution. Normally termed AC current.

**Ampacity:** The current-carrying capacity of an electric device expressed in amperes.

**Ampere:** Unit of electrical current flow. Abbreviated A or amp.

**Apprentice:** Beginner learning a trade. In most states this requires 4 years of on-job training plus school before graduation to journeyman status.

**As-built drawings:** A set of drawings corrected as construction progresses and showing all changes or locations as actually installed during construction.

**Awning window:** A type of window hinged at the top so that the bottom may be opened outward.

**Backfill:** Earth placed against a wall or in an excavation. Required to bring the final surface back to match adjacent surfaces.

**Back splash:** The vertical section of a countertop at the back to prevent spillage down the back wall.

**Ball cock:**   Assembly inside a tank toilet which controls the water supply.

**Balloon frame:**   Wood frame construction in which the studs are continuous from mud sill to top plate. Intermediate floors are supported on ledgers attached to sides of studs.

**Balustrade:**   An assembly of vertical rail supports (balusters) for open stairs.

**Bell-and-spigot:**   Type of connection at the end of cast iron pipe. One end of pipe has an enlarged "hub" or "bell" while the other end is straight. Joint is made with molten lead over compacted oakum.

**Bid:**   The actual proposal from a contractor to an owner to provide labor and materials for construction of a project.

**Bid bond:**   A guarantee that the contractor will not withdraw a bid within a defined period. A guarantee that a low bidder will accept a project contract at the figure quoted.

**Board foot:**   The quantity of wood contained in a piece $12'' \times 12'' \times 1''$. Also termed "board measure," this unit is used to calculate all requirements for lumber used for construction except plywood and certain types of hardwood. Abbreviated BF or BM.

**Boxed rake:**   A cornice at the rake end which is enclosed with fascia and soffit.

**Bracing:**   A general term for strengthening a structure against earthquake, racking, wind damage, or similar loads. Bracing may be obtained by solid plywood, diagonal "let-in" lumber, or by metal straps or rods.

**Branch circuit:**   A separate circuit from the last protection device to the fixture or switch.

**Brick:**   A burned clay product available in many sizes and types used for the construction of walls. "Common" brick are generally used for rough work, while "Normal," "Roman," or other "face" brick are nicer in appearance and are used for exposed work.

**Bridging:**   Lateral bracing of floor joists.

**BTUH:**   Abbreviation for British Thermal Unit per Hour. A Btu is the amount of heat required to raise the temperature of 1 pound of water 1 degree Fahrenheit.

**Built-up roof:**   A type of roof composed of several layers of saturated felt and molten asphalt or coal tar. Usually has a finished surface of a flood coat of liquid into which gravel is embedded.

**Butt joint:**   Type of joint in which one piece of material is connected to another in the same plane and without interlocking parts.

**Butts:**   The construction name for hinges used on doors.

**Casement window:**   A type of window that is hinged at the side and opens outward from the opposite edge.

**Caulking:**   (Calking) Any material used to create a weathertight joint.

**Cement:** A powdered limestone product which is one of the major ingredients for concrete.

**Chord:** The upper or lower member of a truss.

**Circuit breaker:** A protective device installed in an electrical circuit which will automatically "trip" on overload and which may be reset.

**Clean out:** The opening with removable plug provided for clearing clogged sewer lines.

**Closet bend:** A short pipe bend that connects the water closet fixture to the soil line beneath the floor.

**Code:** An ordinance that requires certain compliance for installation or planning of construction.

**Compaction:** The amount or requirement for replacing earth backfill to nearly the original solidity. Usually from about 85 to 95% of original.

**Completion bond:** A bond from the contractor to the owner to assure completion of a project "in accordance with the construction documents."

**Concrete:** A building material composed of cement, aggregate, and water in a homogeneous mass, placed as a plastic material, and hardened by time and curing. Used for foundations, walls, and various other construction. The amounts of each ingredient will allow for the design of different required strengths of finished concrete.

**Conductor:** An electrical wire; a duct. Any material, such as pipe or wire, that will conduct heat or electricity.

**Continuous waste:** A soil line connecting two or more fixtures and using one trap.

**Contours:** Imaginary lines, spaced at specific intervals, which indicate the slope of the ground.

**Contract:** An agreement between two or more parties for work and compensation that is legal and that the law will enforce if necessary.

**Contract documents:** The working drawings, specifications, and other documents used for bidding and construction of a project.

**Corner board:** Vertical boards applied at corners to conceal joints of wood siding.

**Cornice:** A combination of fascia, soffit, and other parts at the lower ends of rafters to provide a finished appearance.

**Course:** A single horizontal row of masonry or roofing.

**Cripple:** Short studs required beneath window framing and at similar locations.

**Damper:** A metal device in a fireplace throat to control the draft. An adjustable metal plate in air-conditioning ducts to control the flow of air. A "fire damper" automatically closes when excess heat melts a thermal connection which normally holds a damper open.

**Dampproofing:** A barrier of water-resistant material applied to walls or floors to control seepage of moisture into a building.

**Datum point:**   Permanent point established for reference above sea level. If such a point is not available, some other reference point may be used as a "datum" for construction.

**Detail:**   A portion of the construction drawings showing the conditions and requirements at a larger scale than is possible on floor plans or elevations.

**Double-hung window:**   A type of window including two sash which slide vertically and may pass each other.

**Drain cock:**   Valve at the lowest point of a water system used to drain the system.

**Drain field:**   The disposal field used in connection with a septic tank.

**Drum trap:**   A cylindrical drum trap with inlet and outlet at different levels.

**Drywall:**   The general term for interior wall finish that is not a "wet" material. Normally includes gypsum board, wood, plywood, or other dry material.

**DWV:**   Common term for "drain, waste, and vent" in plumbing work.

**Elbow:**   A pipe connection which changes the direction of flow in a pipe or duct.

**Elevation:**   The height above sea level of a given point. The small-scale drawings of the exterior views of a building.

**Escutcheon:**   A metal or other decorative ring around a pipe used to conceal the entrance into the wall. The metal plate around a lock or handle on finishing hardware for doors.

**Fascia:**   The horizontal board used to cover the ends of the rafters in a cornice construction.

**Finishing hardware:**   The various locks, butts, closers, push plates, kick plates, and other parts needed to hang a door properly. The hinges and catches for cabinetwork.

**Fitting:**   A general term that includes all types of pipe connections, electrical connections, and similar parts.

**Float:**   A wooden tool used to rough level concrete while still plastic. A "bull float" is a large wooden piece on a long handle which is used to compact and smooth concrete before other work, such as a "troweled finish," is applied.

**Floor plans:**   The small-scale architectural drawings indicating the location and size of the building, walls, windows and doors, and similar parts.

**Flue:**   A vent pipe used to carry off smoke or fumes from heating units.

**Flush door:**   A door constructed with each side smooth and flat. Doors may be hollow-core or solid-core type. Face veneer may be almost anything from manufactured "hardboard" to finest wood veneer.

**Flush valve:** A valve designed to provide a sudden flow of water in a water closet or other plumbing device.

**Footing:** Construction, usually of concrete, beneath a foundation wall to help distribute the imposed loads.

**Formwork:** Wood or metal shapes designed to retain concrete while that material is in a plastic condition.

**Foundation:** The supporting footings, stem wall, and related parts of a structure upon which the remaining portion of the building rest. Usually of concrete or masonry and partially or wholly buried in the earth.

**Frieze:** A horizontal band of wood or other material directly beneath a cornice and above the siding. Often highly ornamented in older construction.

**Fuse:** An electrical safety device designed to melt or "blow" when overloads occur in the circuit which the fuse protects.

**Gable:** The area above the top plate and beneath the rafter line at the ends of a building with a roof slanted in two directions. Also used to indicate a dormer construction projecting from a roof plane.

**Gasket:** A resilient material formed to provide a seal between two metal surfaces.

**Ground:** The earth. Also, a conducting body that serves in place of earth in an electrical system. "To ground" is to provide a pathway for any electrical current.

**Grout:** A special type of thin mortar used between ceramic tiles in place. The cementitious filler used in open cells of concrete unit masonry.

**Gypsum:** A mineral-based material derived from limestone. Used in the manufacture of building boards, plaster of paris, plaster, and similar products.

**Gypsum board:** Termed "gyp board" in building slang. A rigid board manufactured with a solid gypsum core covered both sides and edges with paper. Used extensively instead of plaster for wall surfaces.

**Header:** Horizontal framing members above window and door openings. Same as a lintel.

**Hip rafter:** The rafter running from the ridge to the corner of the building in hip-roof construction.

**Hollow core:** When used in reference to doors, a door manufactured with solid wood rail and stiles and flush faces but with an open grid or spacers for a core stock between the faces.

**Hose bibb:** A water faucet, usually installed outside the building, for attachment of a garden hose. Abbreviated "HB" on drawings.

**Hot wire:** An ungrounded electrical wire, usually with red insulation.

**Insulation:** A material used to reduce the transfer or radiation of heat or cold. A material placed around an electrical wire to prevent accidental grounding. *See* R value *and* U factor.

**Jack rafter:** Roof rafter used between the ridge and valley rafter or between hip rafters and the wall construction.

**Jambs:** Side members of a window or door frame.

**Joist:** Horizontal framing member that supports the floor or ceiling.

**Journeyman:** A skilled workman in the building trades. Usually one who has completed the requirements for apprenticeship.

**Kilowatt:** Unit of electrical power equal to 1000 watts. Abbreviated kw.

**Lally column:** A steel pipe column filled with concrete and used as a beam support.

**Latch set:** A unit of finishing hardware resembling a lock set but without the locking possibility. Used to retain a door shut but not locked.

**Lath:** Wood, gypsum board, or expanded metal used as a base for plaster finish.

**Lien:** A legal action brought against a property, not the owner, by persons who have contributed labor or materials to the project and have not been paid for their services or materials.

**Lineal foot:** A unit of measurement 12″ long used where width or depth is not important.

**Lintel:** Horizontal framing member above window and door openings. Same as a header.

**Lock set:** A unit of finishing hardware for a door. The unit may be a "cylinder lock" with a number of pins that are aligned with a fluted key, or a "box lock" with a "bit" key. May be morticed into the door, set into a series of bored holes, or mounted on the door face.

**Mason:** The journeyman worker who installs brick, concrete masonry units, and similar material.

**Miter joint:** A joint made by cutting the intersecting parts at 45° at a right-angle corner.

**Molding:** A wood or plaster strip used to finish exposed edges. The cross section of moldings may be almost any shape from flat to that with many curves.

**Mortar:** A cement–sand–lime combination with water that is used to bond masonry units together.

**Mullion:** The vertical piece between windows when windows are mounted side by side.

**Muntin:** The dividing strip within a window that separates the various panes of glass.

**Neutral:** The grounding wire of an electrical system that is used to complete the circuit.

**Nipple:** Short length of pipe (under 12″) threaded at each end and used to connect fittings together.

**Nominal size:** The size by which many building products are recognized. For example the nominal size of a 2 × 4 is 2″ × 4″, but the actual size is 1½″ × 3½″ after it is dressed to conform to standards.

**Oakum:** Frayed hemp fibers treated with oil and used to make up joints in bell-and-spigot soil pipe.

**Ohm:** A measure of resistance or impedence for electrical use.

**Outlet:** A general term for any electrical terminal for switches, convenience outlets, or fixtures. The terminal end of an air-conditioning duct. The final end of a sewer or water line.

**Overload:** A condition where a device or member carries more than it was designed for.

**Packing:** Any material used to seal a joint or connection.

**Panel door:** A door composed of solid stiles and rails enclosing thinner panels in a pattern.

**Parging:** A thin coat of cement plaster applied to masonry or concrete walls to obtain a watertight or smooth surface.

**Payment bond:** A bond supplied by the contractor to ensure that all project debts will be paid, thus preventing liens or suits against the owner.

**Penny:** The normal designation of the length of a nail. Usually referred to as "d" or "D."

**Performance bond:** A bond supplied by the contractor to the owner that guarantees that the project "will be completed (performed) in accordance with the construction documents."

**Perimeter drain:** A drain line installed outside the building footing at the perimeter of the building and used to draw off any water that may accumulate against the footing or foundation wall.

**Plaster:** A mixture of cement, sand, lime, and water applied over lath to provide a finished wall or ceiling surface.

**Plastic:** A synthetic compound capable of being molded. In construction plastics are used for pipe, sheet material, protective coverings, and for a great number of molded units.

**Plate:** The horizontal member, usually doubled in residential construction, above the studs. The ceiling joist rests on top of the plate.

**Platform frame:** A system of wood framing in which each story has a set of studs and plates. Multistory buildings are thus built one story on top of the next lower story. Commonly called "Western frame."

**Plot plan:** The drawing indicating all the features of the property and showing the location of the proposed structure and similar details.

**Plywood:**  A sheet material made from multiple odd number of layers of thin veneer glued together with the grain of each layer counter to those on either side. Plywood is stronger, easier to handle and install, and stiffer than solid wood and is extensively used for horizontal diaphragms such as subfloors and roof decks.

**Polyurethane:**  A clear plastic coating material which has excellent durability. Often used in place of varnish.

**Portland cement:**  Powdered limestone ingredient for concrete. Derives its name from the Isle of Portland, England, a limestone deposit area that was originally a good source for natural cement.

**Prime contractor:**  The contractor who has the major agreement with the owner.

**Primer:**  Usually a paint coating applied as a sealer before other coats. A liquid asphalt used to seal porous walls or decks before waterproofing material is installed.

**P trap:**  A plumbing trap in the shape of the letter P but installed as though lying on its side.

**PVC:**  Polyvinyl chloride. A major plastic material used for pipe, sheet, film, and other building products.

**Quantity survey:**  An estimate made by accurately itemizing each material required.

**R value:**  A measurement used to calculate the resistance to the flow of heat. The R value is now marked on almost all insulation material and ranges from about 2.35 to 6.25 per inch of thickness. Refer to Figure 7–1 for recommended values for areas in the United States.

**Rafter:**  The framing members of a roof. Rafters extend from the ridge to the plate line.

**Rake:**  The edge of a roof at the end of a building. The continuation of the cornice applied on a slope at the building end.

**Receptacle:**  A "wall plug" in an electrical installation. A "duplex outlet" has two separate circuits or may be one circuit with the ability to supply two electrical appliances from each of the two openings.

**Resilient flooring:**  Flooring available in squares or rolls, manufactured from asphalt, vinyl, vinyl asbestos, rubber, or similar semisoft material.

**Riser:**  A plumbing "riser" is a vertical run of piping. A stair "riser" is the vertical portion between treads.

**Rough:**  This word is used in connection with plumbing, electrical, framing, and other work to indicate that it is the concealed portion, not the "finished" or exposed portion. It may also be used to indicate that some work is "rough" or not very well installed or finished.

**Sanitary fitting:**  Plumbing fittings that have no inside shoulder that may block the flow of water or sewage.

**Sash:**  The movable portion of a window which contains the glass.

**Saturated felt:** Tough absorbent paper which has been saturated with hot asphalt or coal tar. Usually black in color, this material is used as a moisture barrier beneath wood siding, floors, and for built-up roofs.

**Scale:** Architectural and engineering drawings are drawn at a given ratio to actual size called scale. Scale varies with each drawing but certain drawings are customarily drawn to certain scales. Also, the flat or triangular 12″ ruler used by architects.

**Screed:** A wood or metal edge applied to the wall and used to determine the depth of plaster by using this strip as a depth guide. A straight wooden piece used to level concrete.

**Section:** In architectural drafting, a drawing showing all parts at an imaginary line cut through a wall or other location.

**Seismic force:** An earthquake. Buildings in some areas must be designed to resist seismic action.

**Septic system:** A septic tank, switch boxes, and drainage field used when a municipal sewer system is not available.

**Service entrance:** The general term for the pipe, pot head, main switch, and meter where an electrical system enters the building.

**Shear panel:** A section of a wall covered with plywood or of concrete or masonry which acts as a stiffener in case of earthquake or excessive wind.

**Sheathing:** Often incorrectly pronounced "sheeting". The solid diaphragm of a roof deck or side wall composed of plywood or boards.

**Ship lap:** A type of edge on a wood board which allows a half-lap of one board over the edge of the next board.

**Short circuit:** An accidental connection of low impedance in an electrical curcuit.

**Sill:** The single wood member laid on top of the concrete foundation stem wall. Also called the "mud sill," as this member formerly was often laid directly on the ground or in the "mud."

**Sill cock:** Outside faucet for connection of a hose. A hose bibb.

**Site:** The piece of land upon which a structure is built. The plan for this site includes boundary directions and measurements, contours, location of the structure, and other details and is known as a site plan.

**Sliding window:** A window consisting of two or more sash, one or more that slide horizontally past each other.

**Soffit:** The underside of anything, for example, the soffit of a cornice, the soffit of a beam.

**Soil line:** A plumbing line designed and installed to carry sewage.

**Solid-core door:** A flush-faced door with a solid wood or mineral core. Mineral-cored doors are usually rated as fire doors.

**Span:** The distance between supports of a horizontal beam, floor joist, or rafter.

**Specifications:** The written portions of the construction documents which spell out the requirements for the *quality* of the various materials and the workmanship required. The drawings indicate the *quantity*. The "book" containing the specifications as well as the contract agreements and forms is often referred to as "the specifications" but should be called "the project manual" or a similar name, as the forms and other information do not indicate a quality.

**Square:** In construction terms, a "square" of roofing is an area of 100 sq ft.

**Stack:** In connection with plumbing, a "stack" is the vertical pipe connecting the soil system of various floors and usually extends through the roof as a "vent stack."

**Stool:** The bottom member of a window assembly which forms the sill.

**Stringer:** The inclined members of a stair which support the risers and treads.

**Stucco:** General term for cement plaster used for exterior wall covering. Stucco mesh (chicken wire) is used as a lath base over saturated felt paper.

**Stud:** The vertical framing member resting on the bottom plate and capped with a double top plate. The stud wall construction supports the floor or ceiling construction or the roof construction.

**Subcontractor:** A specialty contractor such as a plumber, electrical contractor, painter, or roofer. A contractor who does not have a "prime" contract directly with the owner but as a "sub" to the prime contractor.

**Subfloor:** The layer of flooring material applied directly to the floor joist. The "finish floor" may be wood, resilient flooring, or many other materials applied over the subfloor.

**Switch:** An electrical device that controls the "on" or "off" function of an electrical fixture or mechanical unit. Also to change direction of flow as in the switch box of a septic system.

**Tee:** A plumbing fitting shaped like a T and with three openings. Any connection that resembles a T.

**Termite:** A soft-bodied wood-eating insect, often with wings, that is common in the southern part of the United States. Aerial termites may swarm similar to bees, but most termites live in the earth (subterranean) and may build mud tunnels or vertical shafts to desirable wood.

**Termite shield:** A sheet metal section installed on top of a concrete foundation and extending over the edges in an effort to prevent termites from building tunnels into the wood above.

**Thermostat:** A switch that is actuated by thermal change. Commonly used for heating and air-conditioning control.

**Toggle bolt:** A bolt equipped with a threaded nut that is designed to upset when pulled on, thereby forming a cross-type anchor in a wall.

**Tongue and groove:**   Milled edge of a wood board or other material in which one edge has a groove and the other edge has a "tongue" which fits into the groove.

**Trap:**   A plumbing device, usually a curved pipe section, which prevents sewer gases from entering the room by blocking them with a water seal. *See* P trap.

**Tread:**   The flat horizontal parts of a stair that are walked upon.

**Trimmer:**   Joist at ends of headers. Studs used to support the headers alongside the studs. Any member used to finish out a window or door frame.

**Truss:**   A built-up unit consisting of upper and lower chords connected with diagonal members. Trusses are usually used in place of solid members when solid stock is not available or becomes too awkward to place or handle.

**U factor:**   Number of Btu transferred in 1 hour by 1 square foot of a building, for each 1°F temperature difference between inside and outside surfaces. The total of all R values of a building assembly.

**Underlayment:**   Any material used over a rough subfloor that will provide a smooth surface for the finish floor.

**Union:**   A plumbing fitting consisting of two threaded ends and a threaded connecting ring. This fitting allows pipe to be connected or dismantled without removing long lengths of pipe.

**Valley:**   The V-shaped connection between two slanted roof planes.

**Valve:**   A plumbing fitting that allows water or gas to be turned off or on at any rate desired.

**Vapor barrier:**   A material used to prevent moisture from transferring from one surface to an adjacent surface. Asphalt-saturated building felt or polyethylene plastic sheet is most commonly used.

**Varnish:**   A clear sealer-finish material composed of resin, oil, thinner, and dryer.

**Vent:**   A plumbing term for the vertical pipe used to aerate the sewer system in a structure.

**Voltage:**   The "pressure" at which an electrical system operates, expressed in volts.

**Water hammer:**   A pounding noise created in water piping by sudden unrelieved surges of water pressure. This noise may be minimized by using an air chamber at pipe ends.

**Watt:**   One volt of electrical pressure times 1 ampere of current equals 1 watt of power. The normal indicator for the amount of power required for lamps and other devices. Abbreviated W.

**Western frame:**   Platform framing. A framing system that provides a complete floor above a stud wall at each story.

**Working drawings:**   The scaled drawings for a structure, including the site plan, foundation plan, floor plan(s), elevations, sections, and details. A major part of the "construction documents," together with the specifications and project manual.

# Appendix

# MEASUREMENT OF AREA AND VOLUMES

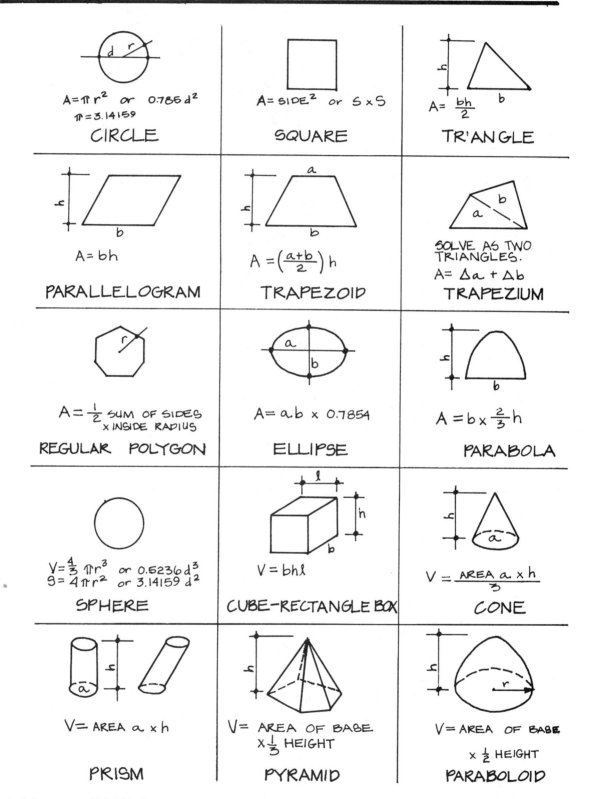

$A = \pi r^2$ or $0.785 d^2$
$\pi = 3.14159$

**CIRCLE**

$A = SIDE^2$ or $S \times S$

**SQUARE**

$A = \dfrac{bh}{2}$

**TRIANGLE**

$A = bh$

**PARALLELOGRAM**

$A = \left(\dfrac{a+b}{2}\right) h$

**TRAPEZOID**

SOLVE AS TWO TRIANGLES.
$A = \Delta a + \Delta b$

**TRAPEZIUM**

$A = \frac{1}{2}$ SUM OF SIDES $\times$ INSIDE RADIUS

**REGULAR POLYGON**

$A = a.b \times 0.7854$

**ELLIPSE**

$A = b \times \dfrac{2}{3} h$

**PARABOLA**

$V = \dfrac{4}{3} \pi r^3$ or $0.5236 d^3$
$S = 4 \pi r^2$ or $3.14159 d^2$

**SPHERE**

$V = bh\ell$

**CUBE-RECTANGLE BOX**

$V = \dfrac{\text{AREA } a \times h}{3}$

**CONE**

$V = \text{AREA } a \times h$

**PRISM**

$V = \text{AREA OF BASE} \times \dfrac{1}{3} \text{HEIGHT}$

**PYRAMID**

$V = \text{AREA OF BASE} \times \frac{1}{2} \text{HEIGHT}$

**PARABOLOID**

*The material on pages 181-89 is from:*
*Jack R. Lewis, ARCHITECTURAL DRAFTSMAN'S REFERENCE HANDBOOK, ©1982.*
*Reprinted by permission of Prentice-Hall, Inc., Englewood Cliffs, N.J.*

# ARCHITECTURAL SYMBOLS

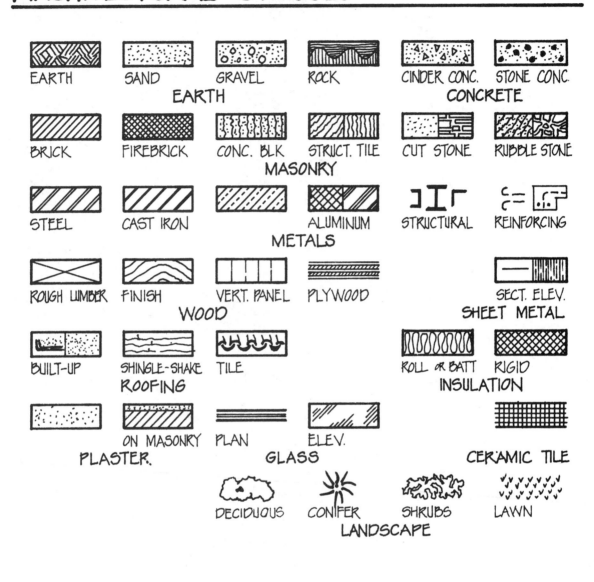

EARTH     SAND     GRAVEL     ROCK     CINDER CONC.    STONE CONC.

EARTH                   CONCRETE

BRICK     FIREBRICK     CONC. BLK     STRUCT. TILE     CUT STONE     RUBBLE STONE

MASONRY

STEEL     CAST IRON           ALUMINUM     STRUCTURAL     REINFORCING

METALS

ROUGH LUMBER   FINISH     VERT. PANEL     PLYWOOD           SECT. ELEV.

WOOD                  SHEET METAL

BUILT-UP     SHINGLE-SHAKE     TILE     ROLL or BATT     RIGID

ROOFING                  INSULATION

ON MASONRY     PLAN     ELEV.

PLASTER.        GLASS              CERAMIC TILE

DECIDUOUS     CONIFER     SHRUBS     LAWN

LANDSCAPE

MASONRY     CONCRETE     MARBLE     SLUMP CONC. BK.

HORIZ. SIDING    VERT. SIDING    PLYWOOD

TYPICAL ELEVATION INDICATION

# MECHANICAL SYMBOLS

—— SOIL LINE    —A— COMRESSED AIR    —ACID— ACID WASTE

—·— COLD WATER    —V— VACUUM    —S— SPRINKLER MAIN

—··— HOT WATER    —+—+ REFRIGERANT    —FO— FUEL OIL

—G— FUEL GAS    —D— DRAIN LINE    —C— CONDENSATE

————— VENT    —CH— CHILLED WATER    —F— FIRE LINE

VALVE, CHECK    AUTOMATIC EXPANSION VALVE

VALVE, DIAPHRAGM    RETURN OR EXHAUST DUCT

VALVE, GATE    SUPPLY DUCT (SHOW SIZE)

VALVE, GLOBE    200 / 100 CFM CEILING AIR OUTLET

VALVE, LOCK AND SHIELD    20 x 12 / 100 CFM WALL AIR OUTLET

VALVE, MOTOR OPERATED    DUCT VOLUME DAMPER

VALVE, PRESSURE REDUCING    LIQUID PUMP

STRAINER    TANK, WATER OR FUEL

CO. CLEAN OUT    COMPRESSOR

HB HOSE BIBB    GAUGE

FLOOR DRAIN    SCALE TRAP

90° ELBOW    WATER CLOSET, TANK TYPE

45° ELBOW    WATER CLOSET, FLUSH VALVE

TEE CONNECTION    LAVATORY OR SINK

CROSS CONNECTION    URINAL

REDUCER    TUB - 4', 5', OR 5'-6"

STOP COCK    TUB, SQ. CORNER TYPE

SHOWER HEAD    SHOWER

183

# ELECTRICAL SYMBOLS

| | | | |
|---|---|---|---|
| ——— | WIRING IN WALL OR CEILING | | SINGLE POLE SWITCH |
| – – – | WIRING IN FLOOR | | THREE-WAY SWITCH |
| - - - - | WIRING EXPOSED | | LOCK OR KEY SWITCH |
| –//– | CONDUIT WITH NUMBER OF WIRES | | SWITCH AND PILOT LIGHT |
| —D | SERVICE WEATHER HEAD | | SWITCH AND DUPLEX RECEPTACLE |
| | STREET LIGHT AND BRACKET | | CEILING PAN |
| △ | TRANSFORMER | | CLOCK RECEPTACLE |
| | PANELBOARD OR MAIN SWITCH | | TELEPHONE |
| | DUPLEX RECEPTACLE | | SIGNAL PUSH BUTTON |
| | TRIPLEX RECEPTACLE | | BUZZER |
| | DUPLEX SPLIT-WIRED | | BELL |
| | SPECIAL-PURPOSE OUTLET | R | RADIO OUTLET |
| | RANGE OUTLET | TV | TELEVISION OUTLET |
| | CEILING LIGHT FIXTURE | | ELECTRIC MOTOR |
| | WALL BRACKET LIGHT FIXTURE | | CIRCUIT BREAKER |
| O | FLUORESCENT LIGHT FIXTURE | | FUSIBLE ELEMENT |
| EXIT | EXIT LIGHT | | NURSE CALL SYSTEM |
| B | BLANKED OUTLET | | PAGING SYSTEM |
| J | JUNCTION BOX | | SOUND SYSTEM |

# NAILS

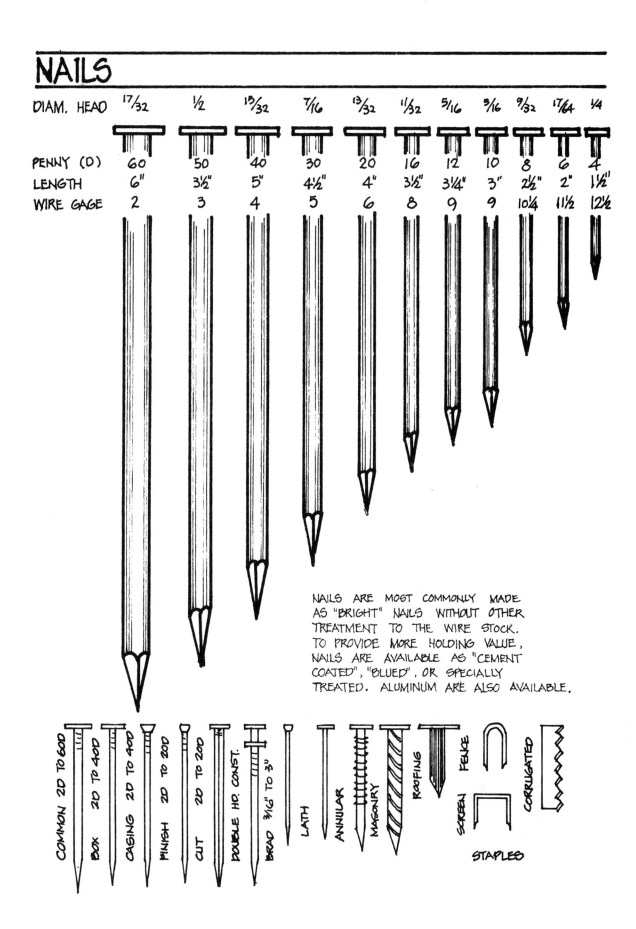

| DIAM. HEAD | 17/32 | 1/2 | 13/32 | 7/16 | 13/32 | 11/32 | 5/16 | 3/16 | 9/32 | 17/64 | 1/4 |
|---|---|---|---|---|---|---|---|---|---|---|---|
| PENNY (D) | 60 | 50 | 40 | 30 | 20 | 16 | 12 | 10 | 8 | 6 | 4 |
| LENGTH | 6" | 3½" | 5" | 4½" | 4" | 3½" | 3¼" | 3" | 2½" | 2" | 1½" |
| WIRE GAGE | 2 | 3 | 4 | 5 | 6 | 8 | 9 | 9 | 10¼ | 11½ | 12½ |

NAILS ARE MOST COMMONLY MADE AS "BRIGHT" NAILS WITHOUT OTHER TREATMENT TO THE WIRE STOCK. TO PROVIDE MORE HOLDING VALUE, NAILS ARE AVAILABLE AS "CEMENT COATED", "BLUED", OR SPECIALLY TREATED. ALUMINUM ARE ALSO AVAILABLE.

COMMON 2D TO 60D
BOX 2D TO 40D
CASING 2D TO 40D
FINISH 2D TO 20D
CUT 2D TO 20D
DOUBLE HD. CONST.
BRAD 3/16" TO 3"
LATH
ANNULAR
MASONRY
ROOFING
SCREEN
FENCE
STAPLES
CORRUGATED

# SCREWS, BOLTS, AND ANCHORS

| | TYPE | LENGTH | GAUGE | | |
|---|---|---|---|---|---|
| **WOOD SCREWS** | FLAT HEAD | ¼ TO 1½ ⅛ INTERVALS | #0 TO #5 | **NUT TYPES** | SQUARE  HEXAGON  CAP  WING |
| | OVAL HEAD | ⅜ TO 3 ¼ INTERVALS | #5 TO #9 | | |
| | ROUND HEAD | ½ TO 3½ ⅝ TO 5 | #9 AND #10 #11 TO #24 | | |
| **SHEET METAL SCREWS** | GIMLET | ¼ TO 4 | #4 TO #14 | **EXPANSION SHIELDS** | MACHINE  BOLTS |
| | BLUNT | | | | |
| | THREADING | | | | LAGS  WOOD SCREW |
| **BOLTS** | MACHINE | 8" TO 30" | ½ TO 1½ | **TOGGLE BOLTS** | SPRING WING |
| | CARRIAGE | 8" TO 20" | ¾ TO 1 | | TUMBLE |
| | FLAT HD. STOVE | 3/16 TO 4 | ⅛ TO ½ | | |
| | RND. HD. STOVE | | | | RIVET |

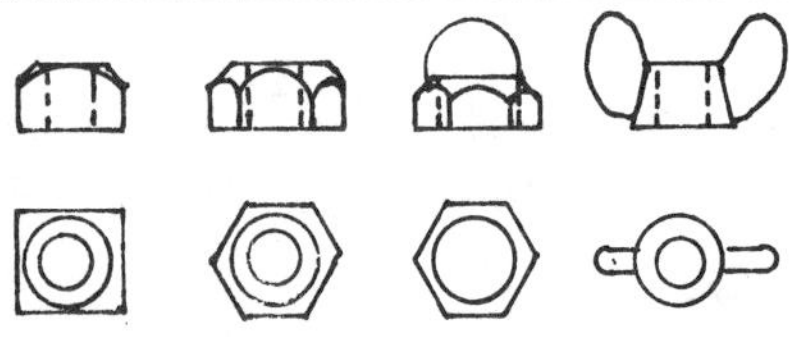

| | | | | | |
|---|---|---|---|---|---|
| **LAGS** | | 1 TO 16 ½ INTERVALS TO 8-1" OVER 8" | ¼ TO 1 | | |

| | GA. | DECIMAL | GA. | DECIMAL |
|---|---|---|---|---|
| **CAP SCREWS** | FILLISTER · HEXAGON · BUTTON | ¾ X 3 TO 2 X 5  ½ X 3½ TO 2 X 6 | | |
| **SET SCREWS** | SQUARE HD. · HEADLESS | ½ TO 5 | ¼ TO 1  #4 TO ½ | |

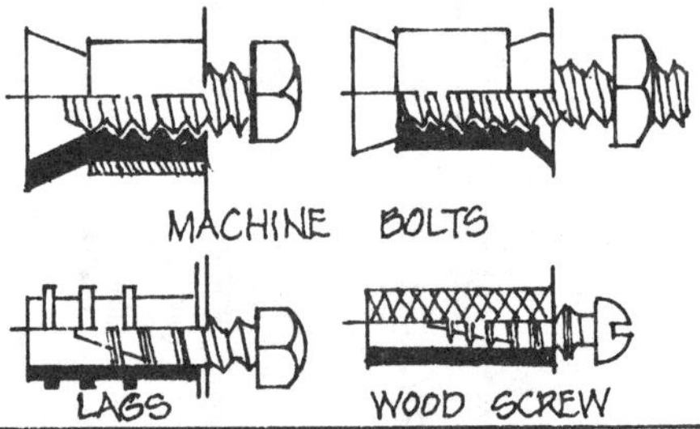

| GA. | DECIMAL | GA. | DECIMAL |
|---|---|---|---|
| #2 | .086 | 5/16" | .313 |
| #3 | .099 | ⅜" | .375 |
| #4 | .112 | 7/16" | .438 |
| #5 | .125 | ½" | .500 |
| #6 | .138 | 9/16" | .563 |
| #8 | .164 | ⅝" | .625 |
| #10 | .190 | ¾" | .750 |
| #12 | .216 | ⅞" | .875 |
| ¼" | .250 | 1" | 1.000 |

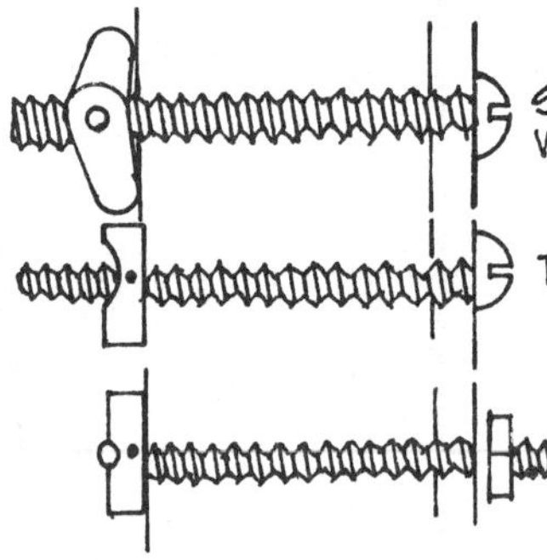

B & W GUAGE FOR SCREWS & BOLTS

186

# LUMBER BOARD MEASURE CONTENTS

| MEMBER SIZE | LENGTH IN FEET | | | | | | | | |
|---|---|---|---|---|---|---|---|---|---|
| | 8 | 10 | 12 | 14 | 16 | 18 | 20 | 22 | 24 |
| 1 X 2 | 1⅓ | 1⅔ | 2 | 2⅓ | 2⅔ | 3 | 3⅓ | 3⅔ | 4 |
| 1 X 3 | 2 | 2½ | 3 | 3½ | 4 | 4½ | 5 | 5½ | 6 |
| 1 X 4 | 2⅔ | 3⅓ | 4 | 4⅔ | 5⅓ | 6 | 6⅔ | 7⅓ | 8 |
| 1 X 6 | 4 | 5 | 6 | 7 | 8 | 9 | 10 | 11 | 12 |
| 1 X 8 | 5⅓ | 6⅔ | 8 | 9⅓ | 10⅔ | 12 | 13⅓ | 14⅔ | 16 |
| 1 X 10 | 6⅔ | 8⅓ | 10 | 11⅔ | 13⅓ | 15 | 16⅔ | 18⅓ | 20 |
| 1 X 12 | 8 | 10 | 12 | 14 | 16 | 18 | 20 | 22 | 24 |
| 1 X 14 | 9⅓ | 11⅔ | 14 | 16⅓ | 18⅔ | 21 | 23⅓ | 25⅔ | 28 |
| 1 X 16 | 10⅔ | 13⅓ | 16 | 18⅔ | 21⅓ | 24 | 26⅔ | 29⅓ | 32 |
| 2 X 4 | 5⅓ | 6⅔ | 8 | 9⅓ | 10⅔ | 12 | 13⅓ | 14⅔ | 16 |
| 2 X 6 | 8 | 10 | 12 | 14 | 16 | 18 | 20 | 22 | 24 |
| 2 X 8 | 10⅔ | 13⅓ | 16 | 18⅔ | 21⅓ | 24 | 26⅔ | 29⅓ | 32 |
| 2 X 10 | 13⅓ | 16⅔ | 20 | 23⅓ | 26⅔ | 30 | 33⅓ | 36⅔ | 40 |
| 2 X 12 | 16 | 20 | 24 | 28 | 32 | 36 | 40 | 44 | 48 |
| 2 X 14 | 18⅔ | 23⅓ | 28 | 32⅔ | 37⅓ | 42 | 46⅔ | 51⅓ | 56 |
| 2 X 16 | 21⅓ | 26⅔ | 32 | 37⅓ | 42⅔ | 48 | 53⅓ | 58⅔ | 64 |
| 3 X 4 | 8 | 10 | 12 | 14 | 16 | 18 | 20 | 22 | 24 |
| 3 X 6 | 12 | 15 | 18 | 21 | 24 | 27 | 30 | 33 | 36 |
| 3 X 8 | 16 | 20 | 24 | 28 | 32 | 36 | 40 | 44 | 48 |
| 3 X 10 | 20 | 25 | 30 | 35 | 40 | 45 | 50 | 55 | 60 |
| 3 X 12 | 24 | 30 | 36 | 42 | 48 | 54 | 60 | 66 | 72 |
| 3 X 14 | 28 | 35 | 42 | 49 | 56 | 63 | 70 | 77 | 84 |
| 3 X 16 | 32 | 40 | 48 | 56 | 64 | 72 | 80 | 88 | 96 |
| 4 X 4 | 10⅔ | 13⅓ | 16 | 18⅔ | 21⅓ | 24 | 26⅔ | 29⅓ | 32 |
| 4 X 6 | 16 | 20 | 24 | 28 | 32 | 36 | 40 | 44 | 48 |
| 4 X 8 | 21⅓ | 26⅔ | 32 | 37⅓ | 42⅔ | 48 | 53⅓ | 58⅔ | 64 |
| 4 X 10 | 26⅔ | 33⅓ | 40 | 46⅔ | 53⅓ | 60 | 66⅔ | 73⅓ | 80 |
| 4 X 12 | 32 | 40 | 48 | 56 | 64 | 72 | 80 | 88 | 96 |
| 4 X 14 | 37⅓ | 46⅔ | 56 | 65⅓ | 74⅔ | 84 | 93⅓ | 102⅔ | 112 |
| 4 X 16 | 42⅔ | 53⅓ | 64 | 74⅔ | 85⅓ | 96 | 106⅔ | 117⅓ | 128 |

ONE BOARD FOOT IS 12" X 12" X 1" NOMINAL. FOR LARGER SIZES THAN GIVEN ABOVE, ADD PROPER COMBINATIONS FOR TOTALS.

# LUMBER SIZES AND PROPERTIES

| NOMINAL SIZE (IN.) b X d | ACTUAL SIZE (IN.) b X d | AREA (IN.²) | MOMENT OF INERTIA (IN⁴) | SECTION MODULUS (IN³) | BOARD MEASURE PER LIN. FOOT | WEIGHT PER LIN. FT. AT 40LB FT³ |
|---|---|---|---|---|---|---|
| 1 X 2 | 3/4 x 1½ | 1.325 | 0.250 | 0.72 | ⅛ | .36 |
| 1 X 3 | 3/4 x 2½ | 1.875 | 0.977 | 0.781 | ¼ | .52 |
| 1 X 4 | 3/4 x 3½ | 2.625 | 2.680 | 1.531 | ⅓ | .73 |
| 1 X 6 | 3/4 x 5½ | 4.125 | 10.398 | 3.781 | ½ | 1.14 |
| 1 X 8 | 3/4 x 7¼ | 5.438 | 23.817 | 6.570 | ⅔ | 1.51 |
| 1 X 10 | 3/4 x 9¼ | 6.938 | 49.466 | 10.695 | ⅚ | 1.93 |
| 1 X 12 | 3/4 x 11¼ | 8.438 | 88.989 | 15.820 | 1 | 2.34 |
| 2 x 2 | 1½ x 1½ | 2.250 | 0.422 | 0.562 | ⅓ | .73 |
| 2 x 3 | 1½ x 2½ | 3.750 | 1.953 | 1.563 | ½ | 1.04 |
| 2 x 4 | 1½ x 3½ | 5.250 | 5.359 | 3.063 | ⅔ | 1.46 |
| 2 x 6 | 1½ x 5½ | 8.250 | 20.797 | 7.563 | 1 | 2.29 |
| 2 x 8 | 1½ x 7¼ | 10.875 | 47.635 | 13.141 | 1⅓ | 3.02 |
| 2 x 10 | 1½ x 9¼ | 13.875 | 98.932 | 21.391 | 1¼ | 3.85 |
| 2 x 12 | 1½ x 11¼ | 16.875 | 177.979 | 31.641 | 2 | 4.69 |
| 2 x 14 | 1½ x 13¼ | 19.875 | 290.775 | 43.891 | 2⅓ | 5.52 |
| 3 x 3 | 2½ x 2½ | 6.250 | 3.250 | 2.600 | ¾ | 1.98 |
| 3 x 4 | 2½ x 3½ | 8.750 | 8.932 | 5.104 | 1 | 2.43 |
| 3 x 6 | 2½ x 5½ | 13.750 | 34.661 | 12.604 | 1½ | 3.82 |
| 3 x 8 | 2½ x 7¼ | 18.125 | 79.391 | 21.901 | 2 | 5.04 |
| 3 x 10 | 2½ x 9¼ | 23.125 | 164.886 | 35.651 | 2½ | 6.42 |
| 3 x 12 | 2½ x 11¼ | 28.125 | 296.631 | 52.734 | 3 | 7.81 |
| 3 x 14 | 2½ x 13¼ | 33.125 | 484.625 | 73.151 | 3½ | 9.20 |
| 3 x 16 | 2½ x 15¼ | 38.125 | 738.870 | 96.901 | 4 | 10.59 |
| 4 x 4 | 3½ x 3½ | 12.250 | 12.505 | 7.146 | 1⅓ | 3.40 |
| 4 x 5 | 3½ x 4½ | 15.750 | 26.580 | 11.811 | 1⅚ | 4.46 |
| 4 x 6 | 3½ x 5½ | 19.250 | 48.526 | 17.646 | 2 | 5.35 |
| 4 x 8 | 3½ x 7¼ | 25.375 | 111.148 | 30.661 | 2⅔ | 7.05 |
| 4 x 10 | 3½ x 9¼ | 32.375 | 230.840 | 49.911 | 3⅓ | 8.93 |
| 4 x 12 | 3½ x 11¼ | 39.375 | 415.283 | 73.828 | 4 | 10.94 |
| 4 x 14 | 3½ x 13½ | 47.250 | 717.609 | 106.313 | 4⅔ | 13.13 |
| 4 x 16 | 3½ x 15½ | 54.250 | 1086.130 | 140.146 | 5⅓ | 15.07 |
| 6 x 6 | 5½ x 5½ | 30.250 | 76.255 | 27.729 | 3 | 8.40 |
| 6 x 8 | 5½ x 7½ | 41.250 | 193.359 | 51.563 | 4 | 11.46 |
| 6 x 10 | 5½ x 9½ | 52.250 | 392.963 | 82.729 | 5 | 14.51 |
| 6 x 12 | 5½ x 11½ | 63.250 | 697.068 | 121.229 | 6 | 17.57 |
| 6 x 14 | 5½ x 13½ | 74.250 | 1127.672 | 167.063 | 7 | 20.63 |
| 6 x 16 | 5½ x 15½ | 65.250 | 1706.776 | 220.229 | 8 | 23.69 |
| 6 x 18 | 5½ x 17½ | 96.250 | 2456.380 | 280.729 | 9 | 26.74 |
| 6 x 20 | 5½ x 19½ | 107.250 | 3398.484 | 348.563 | 10 | 29.79 |
| 6 x 22 | 5½ x 21½ | 118.250 | 4555.086 | 423.729 | 11 | 32.85 |
| 6 x 24 | 5½ x 23½ | 129.250 | 5948.191 | 506.229 | 12 | 35.90 |

# RESILIENT FLOORING

| TYPE | FED. SPEC | SIZE | THICKNESS | USE | BASE | |
|---|---|---|---|---|---|---|
| ASPHALT TILE GREASEPROOF | SS-T-312 SS-T-307 | 9 x 9 | 1/8 | BOS | | A-B-C-D-E GRADES |
| ASBESTOS VINYL TILE | SS-T-312 | 9 x 9, 12 x 12 | 1/16, 3/32, 1/8 | BOS | 2½, 4, 6 | |
| SOLID VINYL TILE CONDUCTIVE | SS-T-312-IV NFPA #56 | 9 x 9, 12 x 12 18 x 18 12 x 12 | 1/16, .080" 3/32, 1/8 1/8 | BOS BOS | | 4 x 36 PLANK 3 x 9 BRICK 4½ x 9 BRICK |
| VINYL SHEET CONDUCTIVE | L-F-001641 L-F-475A NFPA #56 | 6-0, 9-0 12-0 6-0 | .065, .070 .080, .095, .100, .160, .095 | BOS BOS | 2½, 4 AT .08 | |
| RUBBER TILE | SS-T-312 | 9 x 9, 12 x 12, 36 x 36 | 1/8, 3/16, 1/4 | BOS | 1½, 2½, 4, 6 | |
| SHEET RUBBER | ZZ-F-46da | 36 | .093 (3/32), 1/8, 3/16 | BOS | | |
| CORK | LLL-T-00431 | 12 x 12, 6 x 36 | .126 | S | | PREFINISHED OR JOB FIN. |
| CORK CARPET | | 2 METER (78") | 3/16, 1/4 | S | | |
| LINOLEUM TILE | LLL-F-471b | 9 x 9, 12 x 12 | .090 | S | | PRINTED PATTERN |
| LINOLEUM SHEET | LLL-F-1238a | 2 METERS 6-0, 12-0 | .080, .090 1/8 | S | | PRINTED PATTERN |
| PLASTIC | SS-T-312 L-F-475 NFPA #56 | 24 x 24, 36 x 24 24 x 12, 62 x 4 ROLLS 49' x 4' | .080, .100, 1/8, .200 | | | PVC TYPE HEAT-WELD SEAMLESS |

B = BELOW GRADE, O = ON GRADE, S = SUSPENDED

# UNIFORM CONSTRUCTION INDEX—SPECIFICATIONS FORMAT[1]

## DIVISION 1—GENERAL REQUIREMENTS

*Summary of work*
*Alternatives*
*Project meetings*
*Submittals*
*Quality control*
*Temporary facilities and controls*
*Material and equipment*
*Project closeout*

## DIVISION 2—SITE WORK

*Subsurface exploration*
*Clearing*
*Demolition*
*Earthwork*
*Soil treatment*
*Pile foundations*
*Caissons*
*Shoring*
*Site drainage*
*Site utilities*
*Paving and surfacing*
*Site improvements*
*Landscaping*
*Railroad work*
*Marine work*
*Tunneling*

## DIVISION 3—CONCRETE

*Concrete formwork*
*Expansion and contraction joints*
*Concrete reinforcement*
*Cast-in-place concrete*
*Specially finished concrete*
*Specially placed concrete*
*Precast concrete*
*Cementitious decks*

## DIVISION 4—MASONRY

*Mortar*
*Masonry accessories*
*Unit masonry*
*Stone*
*Masonry restoration and cleaning*
*Refractories*

## DIVISION 5—METALS

*Structural metal framing*
*Metal joists*
*Metal decking*
*Lightgage metal framing*
*Metal fabrications*
*Ornamental metal*
*Expansion control*

## DIVISION 6—WOOD AND PLASTICS

*Rough carpentry*
*Heavy timber construction*
*Trestles*
*Prefabricated structural wood*
*Finish carpentry*
*Wood treatment*
*Architectural woodwork*
*Prefabricated structural plastics*
*Plastic fabrications*

## DIVISION 7—THERMAL AND MOISTURE PROTECTION

*Waterproofing*
*Dampproofing*
*Insulation*
*Shingles and roofing tiles*
*Preformed roofing and siding*
*Membrane roofing*
*Traffic topping*
*Flashing and sheet metal*
*Roof accessories*
*Sealants*

## DIVISION 8—DOORS AND WINDOWS

*Metal doors and frames*
*Wood and plastic doors*
*Special doors*
*Entrances and storefronts*
*Metal windows*
*Wood and plastic windows*
*Special windows*
*Hardware and specialties*
*Glazing*
*Window walls/curtain walls*

[1]Courtesy of Construction Specifications Institute.

## DIVISION 9—FINISHES

*Lath and plaster*
*Gypsum wallboard*
*Tile*
*Terrazzo*
*Acoustical treatment*
*Ceiling suspension systems*
*Wood flooring*
*Carpeting*
*Special flooring*
*Floor treatment*
*Special coatings*
*Painting*
*Wall covering*

## DIVISION 10—SPECIALITIES

*Chalkboards and tackboards*
*Compartments and cubicles*
*Louvers and vents*
*Grilles and screens*
*Wall and corner guards*
*Access flooring*
*Specialty modules*
*Pest control*
*Fireplaces*
*Flagpoles*
*Identifying devices*
*Pedestrial control devices*
*Lockers*
*Protective covers*
*Postal specialties*
*Partitions*
*Scales*
*Storage shelving*
*Sun control devices (exterior)*
*Telephone enclosures*
*Toilet and bath accessories*
*Wardrobe specialties*

## DIVISION 11—EQUIPMENT

*Built-in maintenance equipment*
*Bank and vault equipment*
*Commercial equipment*
*Checkroom equipment*
*Darkroom equipment*
*Ecclesiastical equipment*
*Educational equipment*
*Food service equipment*
*Vending equipment*
*Athletic equipment*
*Industrial equipment*

## DIVISION 11—EQUIPMENT (CONT'D)

*Laboratory equipment*
*Laundry equipment*
*Library equipment*
*Medical equipment*
*Mortuary equipment*
*Musical equipment*
*Parking equipment*
*Waste handling equipment*
*Loading dock equipment*
*Detention equipment*
*Residential equipment*
*Theater equipment*
*Registration equipment*

## DIVISION 12—FURNISHINGS

*Artwork*
*Cabinets and storage*
*Window treatment*
*Fabrics*
*Furniture*
*Rugs and mats*
*Seating*
*Furnishing accessories*

## DIVISION 13—SPECIAL CONSTRUCTION

*Air supported structures*
*Integrated assemblies*
*Audiometric room*
*Clean room*
*Hyperbaric room*
*Incinerators*
*Instrumentation*
*Insulated room*
*Integrated ceiling*
*Nuclear reactors*
*Observatory*
*Prefabricated buildings*
*Special purpose books and buildings*
*Radiation protection*
*Sound and vibration control*
*Vaults*
*Swimming pool*

## DIVISION 14—CONVEYING SYSTEMS

*Dumbwaiters*
*Elevators*
*Hoists and cranes*
*Lifts*

DIVISION 14—CONVEYING SYSTEMS
(CONT'D)

*Material handling systems*
*Turntables*
*Moving stairs and walks*
*Pneumatic tube systems*
*Powered scaffolding*

DIVISION 15—MECHANICAL

*General provisions*
*Basic materials and methods*
*Insulation*
*Water supply and treatment*
*Waste water disposal and treatment*
*Plumbing*
*Fire protection*
*Power or heat generation*
*Refrigeration*

DIVISION 15—MECHANICAL
(CONT'D)

*Liquid heat transfer*
*Air distribution*
*Controls and instrumentation*

DIVISION 16—ELECTRICAL

*General provisions*
*Basic materials and methods*
*Power generation*
*Power transmission*
*Service and distribution*
*Lighting*
*Special systems*
*Communications*
*Heating and cooling*
*Controls and instrumentation*

# Practice
# Plans

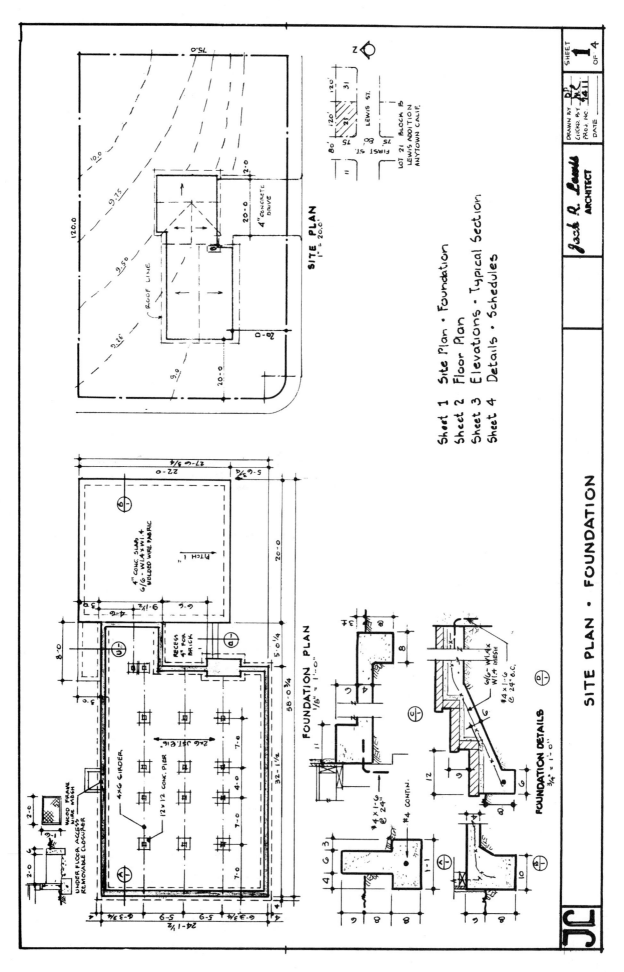

SITE PLAN • FOUNDATION

Sheet 1  Site Plan • Foundation
Sheet 2  Floor Plan
Sheet 3  Elevations • Typical Section
Sheet 4  Details • Schedules

SITE PLAN
1" = 20'-0"

FOUNDATION PLAN
1/8" = 1'-0"

FOUNDATION DETAILS
3/4" = 1'-0"

Jack R. Lewis
ARCHITECT

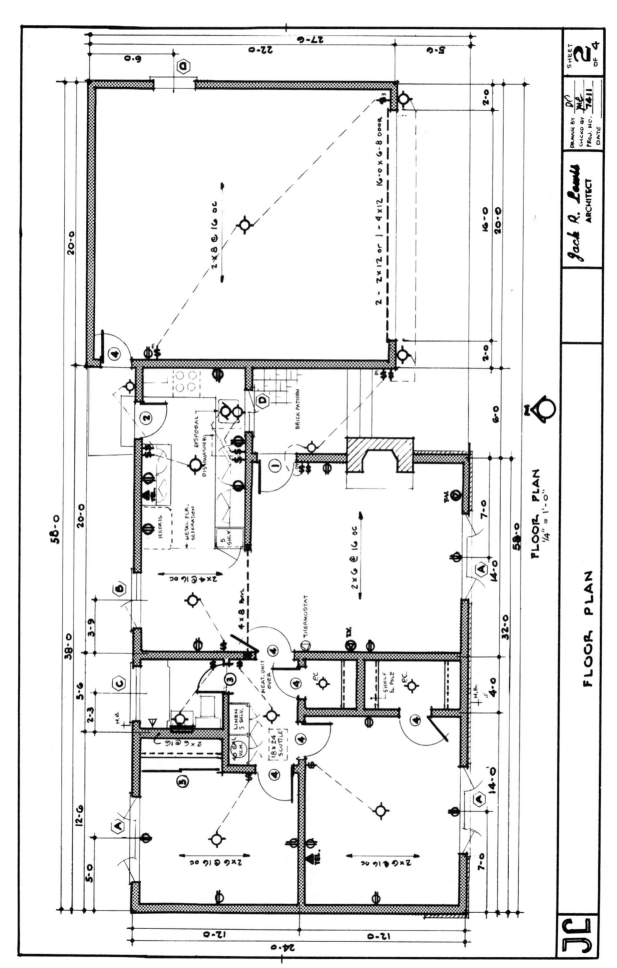

FLOOR PLAN

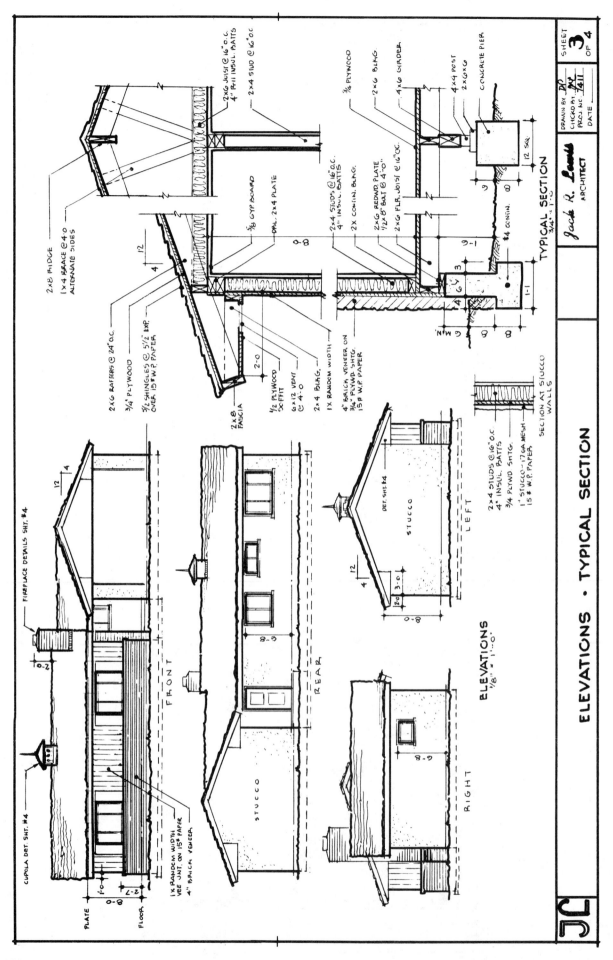

ELEVATIONS · TYPICAL SECTION

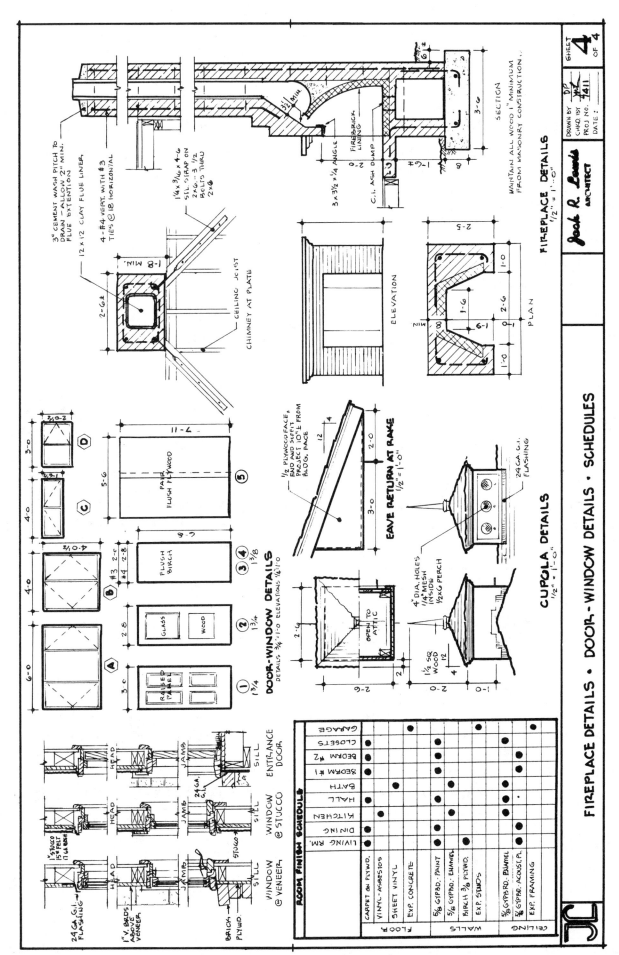

FIREPLACE DETAILS • DOOR-WINDOW DETAILS • SCHEDULES

# RECOMMENDED REFERENCES

## DESIGN AND DRAFTING

BELLIS, HERBERT F. and SCHMIDT, WALTER A., *Architectural Drafting.* New York: McGraw-Hill, 1971.

BENJAMIN, B. S., *Building Construction for Architects and Engineers,* Vol. 2. Lawrence, KS: Ashnorjen Bezaleel Publishing Company, 1979.

HEPLER, DONALD E. and WALLACH, PAUL I., *Architect, Drafting and Design.* New York: McGraw-Hill, 1965.

MCGINTY, TIM, *Drawing Skills in Architecture.* Dubuque, Iowa: Kendall/Hunt, 1976.

MULLER, EDWARD J., *Architectural Drawing.* Englewood Cliffs, NJ: Prentice-Hall, 1967.

MULLER, EDWARD J., *Reading Architectural Working Drawings,* 2nd ed. Englewood Cliffs, NJ: Prentice-Hall, 1971.

PATTEN, L. M. and ROGNESS, M. L., *Architectural Drafting.* Dubuque, Iowa: Kendall/Hunt, 1962.

WEIDHAAS, ERNEST R., *Architectural Drafting and Construction.* Boston: Allyn and Bacon, 1972.

## SPECIFICATIONS

LEWIS, JACK R., *Construction Specifications.* Englewood Cliffs, NJ: Prentice-Hall, 1975.

ROSEN, HAROLD J., *Principles of Specification Writing.* New York: Reinhold, 1967.

## ESTIMATING

COOPER, GEORGE H., *Building Construction Estimating,* 2nd ed. New York: McGraw-Hill, 1959.

DAGOSTINO, FRANK R., *Estimating in Building Construction,* 2nd ed. Reston, VA: Reston, 1978.

FOSTER, NORMAN, *Construction Estimates From Take-off to Bid.* New York: McGraw-Hill, 1961.

HORNUNG, WILLIAM J., *Estimating Building Construction.* Englewood Cliffs, NJ: Prentice Hall, 1970.

PULVER, H. E., *Construction Estimates and Costs.* New York: McGraw-Hill, 1960.

STEINBERG, JOSEPH and STEMPEL, MARTIN, *Estimating for Building Trades.* Chicago: American Technical Society, 1965.

VAN ORMAN, HALSEY, *Estimating for Residential Construction.* Albany, NY: Delmar Publishers, 1978.

## MISCELLANEOUS

BALL, JOHN E., *Light Construction Techniques.* Reston, VA: Reston, 1980.

KETCHUM, MILO S., *Handbook of Standard Details for Buildings.* Englewood Cliffs, NJ: Prentice-Hall, 1956.

LEWIS, JACK R., *Architectural Draftsman's Reference Book.* Englewood Cliffs, NJ: Prentice-Hall, 1982.

REINER, LAURENCE E., *Methods and Materials of Residential Construction.* Englewood Cliffs, NJ: Prentice-Hall, 1981.

# Index